T0275656

Mass Spectrometry

Mass Spectrometry
Techniques for Structural Characterization of Glycans

Michael A. Madson
BioLogistics LLC
Ames, IA, USA

ELSEVIER

AMSTERDAM • BOSTON • HEIDELBERG • LONDON • NEW YORK • OXFORD
PARIS • SAN DIEGO • SAN FRANCISCO • SINGAPORE • SYDNEY • TOKYO

Elsevier
Radarweg 29, PO Box 211, 1000 AE Amsterdam, Netherlands
The Boulevard, Langford Lane, Kidlington, Oxford OX5 1GB, UK
50 Hampshire Street, 5th Floor, Cambridge, MA 02139, USA

Notices
Knowledge and best practice in this field are constantly changing. As new research and
experience broaden our understanding, changes in research methods or professional practices,
may become necessary.

Practitioners and researchers must always rely on their own experience and knowledge in
evaluating and using any information or methods described herein. In using such information
or methods they should be mindful of their own safety and the safety of others, including
parties for whom they have a professional responsibility.

To the fullest extent of the law, neither the Publisher nor the authors, contributors, or editors,
assume any liability for any injury and/or damage to persons or property as a matter of products
liability, negligence or otherwise, or from any use or operation of any methods, products,
instructions, or ideas contained in the material herein.

British Library Cataloguing-in-Publication Data
A catalogue record for this book is available from the British Library

Library of Congress Cataloging-in-Publication Data
A catalog record for this book is available from the Library of Congress

ISBN: 978-0-12-804129-1

For Information on all Elsevier publications
visit our website at https://www.elsevier.com/

Working together
to grow libraries in
developing countries

www.elsevier.com • www.bookaid.org

DEDICATION

We dedicate this book to my Lord and Savior Jesus the Christ, His Father & mine and the Holy Spirit who have spent countless hours with me in writing this book so that many others may grasp the power of these techniques and apply them to their quests in science. We hope chemistry and medicine will be advanced and lives will be saved or lengthened because of Their intervention into my life.

CONTENTS

Mass Spectral Analysis of Carbohydrates

1 ISOLATION METHODS FOR MS ANALYSIS

In this book, we describe the methods required for isolation of intact oligosaccharides for analysis by mass spectrometry. There are two basic methods we use. One is done by passing ammonium salts through a cation exchange cartridge in the ammonium form [1,2], the second through ethanol extraction and removal of the centrifuge samples' supernatant. Both methods remove protein, by cationic adsorption of the protein in [3] the first case and protein precipitation in the second case. We have examples of both methods and one example of both methods [3] used on the same glycan [3]. The use of ammonium salts through cationic exchange resin in the ammonium form requires careful handling which sometimes may be difficult to accomplish [1,2]. The method is described below.

2 COMPONENT MONOSACCHARIDE ANALYSIS BY ACID CATALYZED HYDROLYSIS OF GLYCAN AND QUALITATIVE CHROMATOGRAPHIC IDENTIFICATION BY HPLC

The monosaccharide composition of the glycan to be analyzed is accomplished by known methods. Generally, the glycan to be analyzed is hydrolyzed with strong acid by HCl solution at various concentrations of acid, various times and various temperatures due to the varied acid labilities and stabilities of the oligosaccharides and monosaccharides to be analyzed. Two different hydrolytic conditions [4–6] are used for the monosaccharide composition of bovine thyroglobulin O-linked oligosaccharide [4–6] and are shown below. The resulting High Performance Anion Exchange Chromatography-Pulsed Amperometric Detection (HPAEC-PAD) chromatograms are shown also [1,2].

3 IDENTIFICATION OF GLYCOSES VERSUS GLYCITOLS BY HPAEC-PAD DIFFERENCE CHROMATOGRAMS

Often glycans can be in the reducing sugar or glycitol form. An example of glycan reducing sugar analysis and the corresponding glycan alditol analysis can be done and is shown below [4–6]. This work establishes which monosaccharide is the reducing sugar of the glycan. That is, the reducing glycan is hydrolyzed, possibly using more than one hydrolytic conditions, to give the complete monosaccharide composition. Then the reducing glycan is reduced with $NaBH_4$ with standard conditions (given below) and the reducing monosaccharide will disappear from the monosaccharide composition, the alditol will have a different retention time. Most of the time monosaccharide alditols have a much lower retention time than the corresponding reducing monosaccharides. An example is given for bovine thyroglobulin O-linked oligosaccharide.

4 THE METHODS FOR PREPARING AN OLIGOSACCHARIDE REDUCING GLYCAN FROM A GLYCOPROTEIN, LINKED THROUGH O-GLYCOSIDIC LINKAGE

To help identify the reducing sugar of a glycan, as noted above, to provide a reactive site for further derivatization to allow further characterization by MS or chromatography, a method for the nonreductive β-elimination of O-linked oligosaccharides from glycoprotein has been developed. The method involves first removing any N-linked oligosaccharides by the method described (N- and O-linked oligosaccharide isolation method) [1,2,4–7]. The O-linked oligosaccharide, removed from the ammonium form cation exchange cartridge is treated, as shown below, to release the glycan which was formerly linked to protein by an O-glycosidic linkage.

4.1 MS: Electrospray Ionization Mass Spectrometry of Glycans

As an example of negative ion ESI-MS of glycans, we chose banana fruit extract ESI-MS [8–19]. The oligosaccharide was unknown. We extracted banana fruit with 95% ethanol, directly, and used the method above for isolation of the oligosaccharide. The fourth spectrum from the top is from the ethanol extract of banana fruit (Fig. 4.1).

The spectrum base peak is m/z 283. Its proposed ion structure is shown in Fig. 4.2.

To propose the structure noted above we use the following approach: The monosaccharide composition is determined as noted above. In this case we assumed it to be a kestose found in other

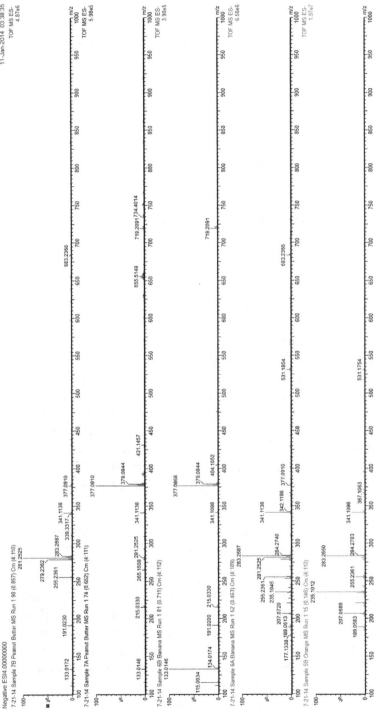

Figure 4.1

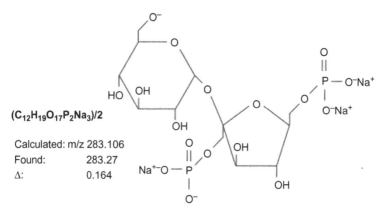

$(C_{12}H_{19}O_{17}P_2Na_3)/2$

Calculated: m/z 283.106
Found: 283.27
Δ: 0.164

Figure 4.2 This ion (negative mode MS) is a cleavage product of the trisaccharide fructosyl sucrose 1,6-diphosphate, neokestose-1, 6-diphosphate, from the ethanol extract of banana. It is the sucrose portion of the molecule. Here the cleavage is shown to be at the O-6' position. This molecule is in the form of the tetraanion diphosphate. There are three sodium ions and only one unneutralized anion. This is a common phenomenon whereby one or more of the anion—cation pair is recognized as an anion. In this case having a net minus 2 charge.

previous works in the literature [9,13,14,20–23]. We knew it contained phosphate or sulphate through a described method [23] for the discernment between phosphate or sulfate substitution. We devise a combination of data, to find an ion that fits the data, that is, charge data, monosaccharide composition, and reducing sugar data to include the MS data. With these possible structures we need to calculate the ion masses for each one. We use the rule that the difference between calculated and found masses must be within the acceptable accuracy of the instrument used. In this case the instrument was an inductively coupled plasma electrospray ionization mass spectrometry (ICP ESI MS). This instrument usually gives a less than 0.3 amu difference between calculated and found ionic mass.

Also, for confirmation of the ion, we carry out MS/MS of the ion studied as in the case of m/z 283 (Figs. 4.3 and 4.4) [8,24].

The MS/MS spectrum of m/z 283, is the fourth spectrum from the top. It shows the m/z 341 peak as part of its MS/MS spectrum. We find the ion structure of the daughter ion by the same method as with the parent ion, m/z 283. This means that the parent ion m/z 283, must have the sucrosyl portion of the ion. This means that the two phosphate groups have been hydrolyzed in the mass spectrometer. This is consistent with the parent ion being from neokestose-1,6-diphosphate [8].

$C_{12}H_{21}O_{11}$

Calculated: m/z   341.2604
Found: 341.1078
Δ: 0.1526 amu

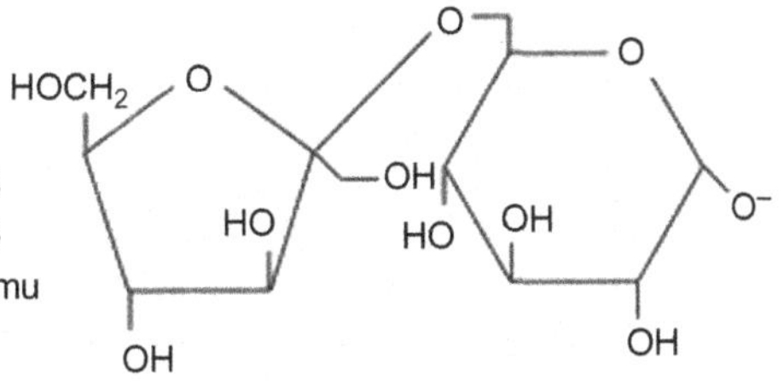

HOCH$_2$

Figure 4.3

4.2 Hydride Insertion Addition Reaction for the Derivatization of Phosphorylated or Sulfated Glycans and the Discernment of Phosphate Versus Sulfate Ester Substitution MS Method

We have discovered a reaction to "label" phosphate and sulfate carbohydrate esters [25–31]. This derivatization lends stability to the phosphorylated or sulfated esters. For the phosphate ester one hydride anion adds to the phosphoryl group and two hydride anions add to the two sulfuryl groups. Not only does this reaction give a confirmatory set of MS and MS/MS ions for the same compound, it also provides a way to discern between sulfate versus phosphate carbohydrate esters. The latter happens because they will have a difference between sulfated and phosphorylated structures of 2.0 amu for a singly charged anion and 4.0 for doubly phosphorylated and doubly sulfated anions.

For an example of the reaction we turn to our work with an unknown bovine milk trisaccharide [4–6,10–12,20,32–35]. The following atmospheric pressure ionization-mass spectrometry (API-MS) for bovine milk oligosaccharide shows two main ions, m/z 377 and m/z 439. With MS/MS of the underivatized molecule we obtain m/z 97.0 and m/z 78.9 for both the m/z 377 and m/z 439 parent ions. The m/z 97.0 is indicative of sulfate and phosphate but the m/z 78.9 is diagnostic for phosphate only. Thus we can conclude that these parent ions are both phosphorylated. We also performed the hydride addition reaction on the milk trisaccharide and found an m/z 369 ion only in its MS spectrum (Fig. 4.5).

Fig. 4.5 is the MS spectrum for the bovine milk oligosaccharide. Note the ions m/z 377 and m/z 439 (Fig. 4.6).

This mass spectrum is from the milk oligosaccharide extracted from milk with 95% ethanol. The structure for the base peak, m/z 683.4, is a cross-ring cleavage product between C-2 and O-5 of the glucose

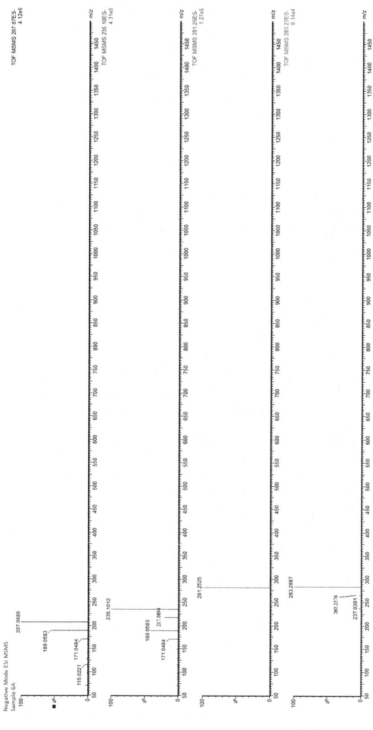

Figure 4.4

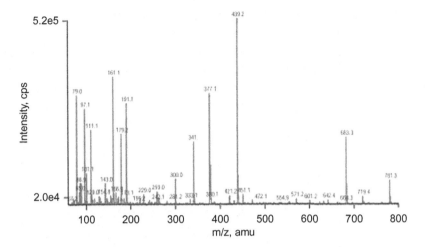

Figure 4.5

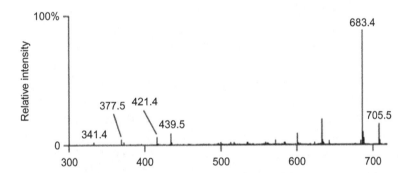

Figure 4.6

portion of 3'-*N*-acetyl neuraminyl lactose-6'-phosphate [20]. This ion, as well as m/z 341.4, m/z 377.5, m/z 421.4 and m/z 439.5 are all present in the mass spectrum of the product of the ammonium form cation exchange resin method for the isolation of this trisaccharide from fat free milk (Fig. 4.7).

The structure for the m/z 439 ion and the structure for m/z 377 are below. The MS/MS spectrum above, ion, m/z 79.0, is evidence for phosphate ester substitution of the m/z 439 ion (Figs. 4.8 and 4.9).

In the following chromatogram of unhydrolyzed milk olgiozaccharide, analyzed less than or equal to 2 minutes at room temperature we have a single peak with no lactose in this mixture. But when allowed

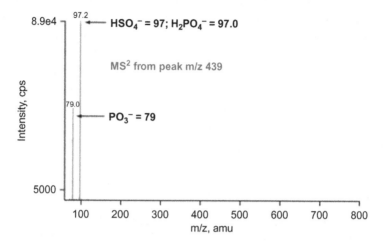

Figure 4.7

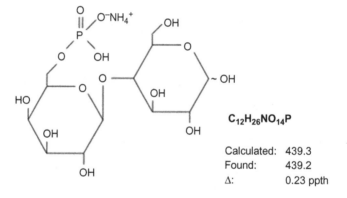

Figure 4.8

to stand 10 minutes at room temperature, either in the free acid form or ammonium ion form it decomposes to lactose (see Fig. 4.10).

The milk oligosaccharide was acid hydrolyzed in the usual way [33] with 1 N HCl for 1 hour at 100°C. Its monosaccharide components are shown in Fig. 4.11.

Hydrolysis of unreduced milk trisaccharide preparation. These are the neutral sugars from 3'-sialyl lactose-6'-phosphate on a Dionex CarboPac PA 20 column. After hydrolysis neuraminic acid and phosphate would not chromatograph at appreciable retention times in this high performance liquid chromatography (HPLC) program. Nor is

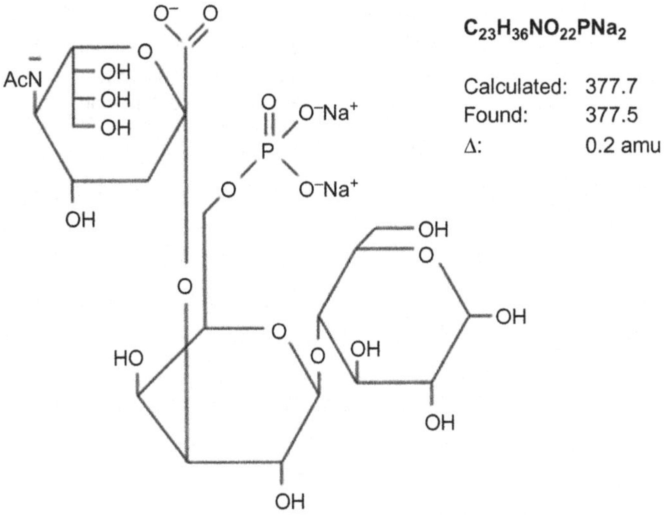

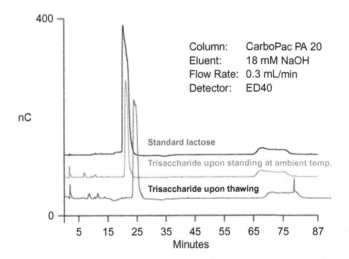

$C_{23}H_{36}NO_{22}PNa_2$

Calculated: 377.7
Found: 377.5
Δ: 0.2 amu

Figure 4.9

Figure 4.10

there any glucosamine. If there were glucosamine found here, it would indicate that the molecule found might be the known milk oligosaccharide, sialyl lactosamine phosphate. The chromatogram from the reduction and hydrolysis of this molecule (not shown) has only galactose with some contaminating glucose, indicating that [20] glucose is the reducing sugar [36,37].

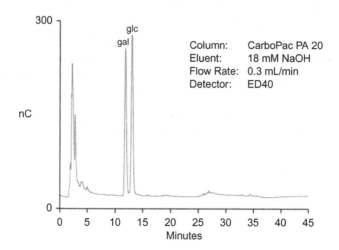

Figure 4.11

4.3 Hydride Insertion Reduction Reaction Protocol
4.3.1 Commercial Fat Free Milk (0.1 mL)
1. Dilute to 1.0 mL with water.
2. Push through 1 mL ammonium form ion exchange cartridge.
3. Freeze dry to ~0.1 mL.
4. Make up to 1.0 mL with water.
5. Freeze.
6. Allow to thaw, partially.
7. Use 0.05 mL for reduction reaction.

4.3.2 3′-Sialyl Lactose-6′-phosphate/Sulfate
1. Make up to 1.0 mL with pH 11.4 ammonium hydroxide, immediately.
2. Add sodium borohydride (4 N, 3 μL).
3. Solution bubbles immediately.
4. Let stand at ambient temperature for 1 hour.
5. Immediately freeze dry to ~0.1 mL.
6. Dissolve in 1.0 mL water.
7. Freeze. Partially thaw and analyze within 2 minutes by ESI negative ion triple quadrapole MS.

4.3.3 Reduced 3′-Sialyl Lactose-6′-phosphate/Sulfate
The MS spectrum for the hydride added bovine milk oligosaccharide is Fig. 4.12.

In Fig. 4.12 there is one main peak indicating one main component from the bovine milk oligosaccharide. Its structure is as shown in the following figure (Fig. 4.13).

Another example of the [4–6,25–31] use of MS/MS and the hydride addition derivatization reaction comes from the ethanol

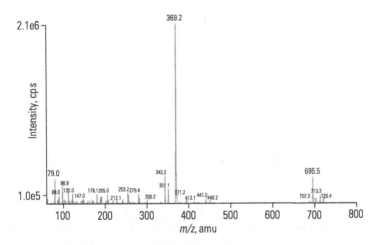

Figure 4.12 This Figure results from the protocol that allows phosphate/sulfate discernment of the phosphate or sulfate carbohydrate ester from the bovine milk trisaccharide. This base peak is the M-2H/2 ion, m/z 369.2. Its proposed structure is shown in Figure 16.

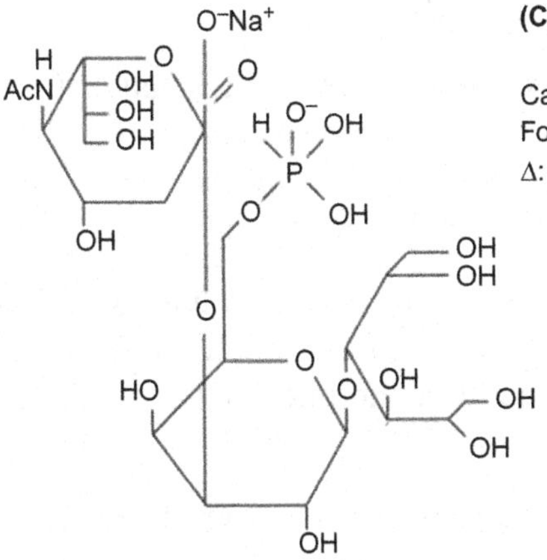

$(C_{23}H_{42}NO_{22}PNa)/2$

Calculated: 369.2
Found: 369.2
Δ: 0 amu

Figure 4.13

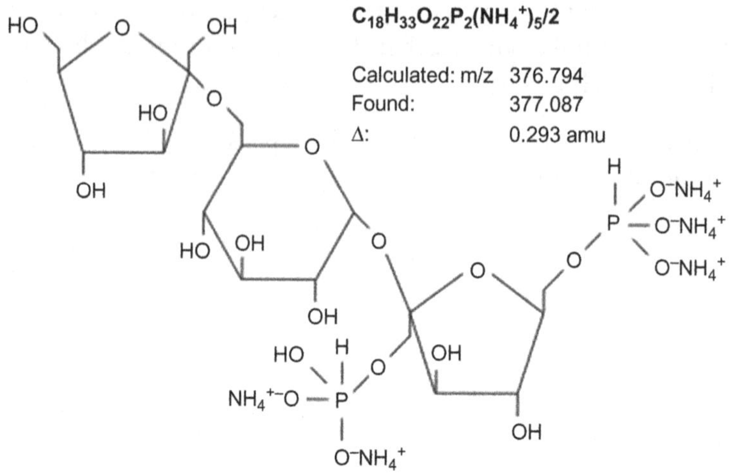

Figure 4.14

extract of banana fruit. The m/z 377 peak could be the ion shown in the following figure (Fig. 4.14).

We have a sulfated O-linked oligosaccharide from O-linked bovine thyroglobulin which is drawn much like the milk oligosaccharide above. Like our current bovine thyroglobulin oligosaccharide, the sulfate/phosphate discernment method is used for this milk oligosaccharide. Here we find that the milk oligosaccharide is phosphorylated and not sulfated. The O-linked oligosaccharide from bovine thyroglobulin is, on the other hand, sulfated. These conclusions are supported by MS/MS data for both compounds (Fig. 4.15).

Here we have the whole trisaccharide diphosphate. It is neokestose-1,6-diphosphate. Its MS/MS shows its own peak, m/z 377, and m/z 341. The m/z 341 structure suggests that the trisaccharide is cleaved between the C-2 and O-1' of the molecule. It is shown in Fig. 4.16.

The determination of phosphate versus sulfate substitution of the neokestose-diphosphate or neokestose-disulfate is shown in Fig. 4.17 by comparing the two structures, the phosphate or sulfate substituted structures [8–19].

Here we have the dianion, noted above, which is 1.723 amu difference between calculated and found molecular masses. In this case we see that the molecule is not disulfated. And in the following slide we

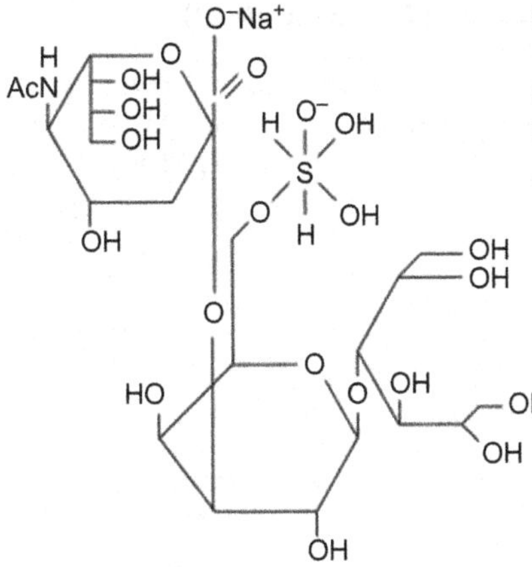

$(C_{23}H_{43}NO_{22}SNa)/2$

Calculated: 370.3
Found: 369.2
Δ: 1.1 amu

Figure 4.15

$C_{12}H_{21}O_{11}$

Calculated: m/z 341.2604
Found: 341.1078
Δ: 0.1526 amu

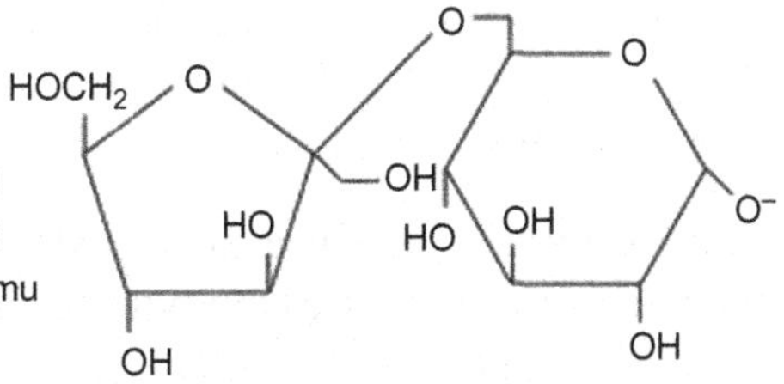

Figure 4.16

$(C_{18}H_{35}O_{22}S_2(NH_4^+)_5/2$

Calculated: m/z 378.81
Found: 377.088
Δ: 1.723 amu

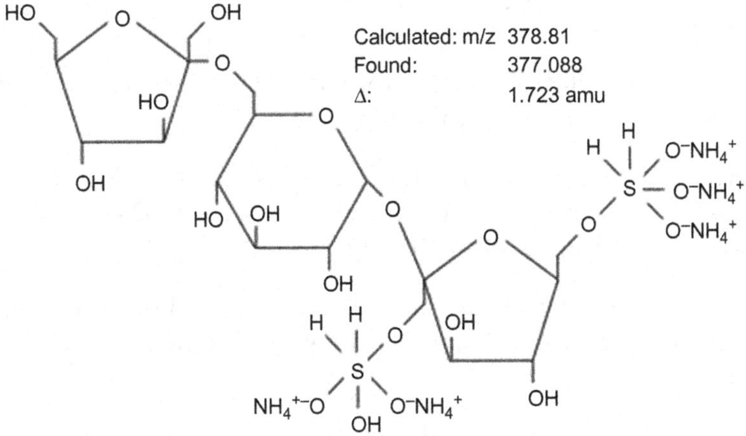

Figure 4.17

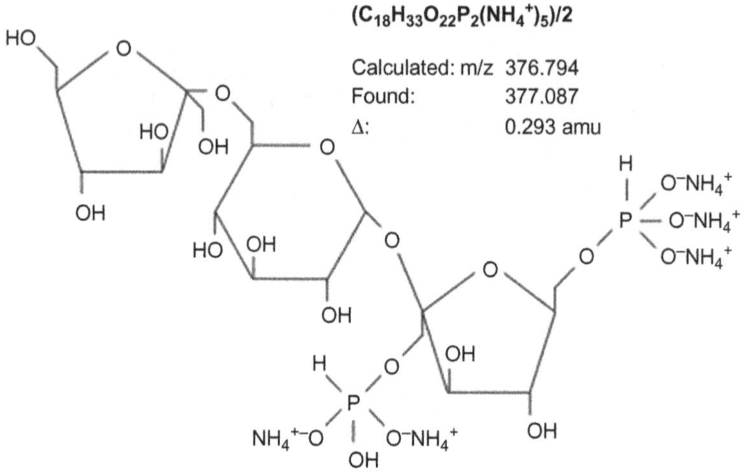

$(C_{18}H_{33}O_{22}P_2(NH_4^+)_5)/2$

Calculated: m/z 376.794
Found: 377.087
Δ: 0.293 amu

Figure 4.18

find a difference between calculated and found ionic mass for the diphosphorylated ion is 0.293 amu, well within the accuracy of the ICP ESI MS instrument (Fig. 4.18).

Another example of the use of MS/MS and hydride addition reaction for the identification of carbohydrates, is the following ionic structure for m/z 719 MS (Figs. 4.19–4.22).

The previous three figures from the MS/MS of m/z 719 are m/z 377, m/z 341 and m/z 215. Note the differences between the calculated and found molecular masses are well within instrumental accuracy (Fig. 4.23).

This is the disulfated structure for m/z 723.5598 whereas the diphosphorylated structure is shown in Fig. 4.24.

Here is the diphosphorylated drawn structure. The difference between the calculated and found ionic masses is 0.32 amu, which is well within the accuracy of the ICP-ESI MS.

There is direct evidence of the hydride addition reaction in the structure for m/z 133 ion. Its drawn structure is shown [25,26] (Fig. 4.25).

Its MS/MS shows itself and m/z 115 as the daughter ions, found in Fig. 4.26.

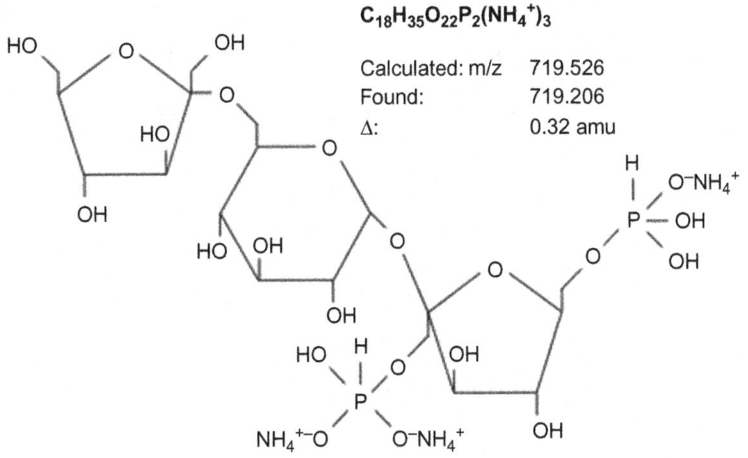

$$C_{18}H_{35}O_{22}P_2(NH_4^+)_3$$

Calculated: m/z   719.526
Found: 719.206
Δ: 0.32 amu

Figure 4.19

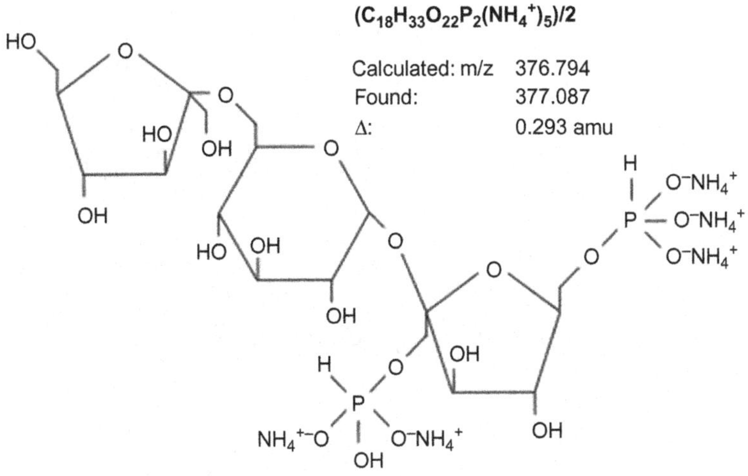

$$(C_{18}H_{33}O_{22}P_2(NH_4^+)_5)/2$$

Calculated: m/z   376.794
Found: 377.087
Δ: 0.293 amu

Figure 4.20

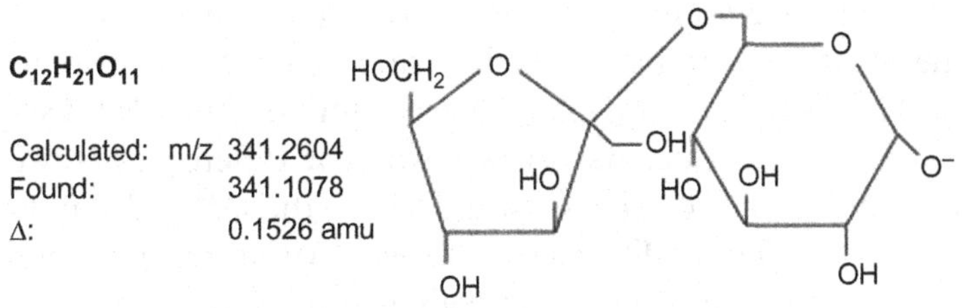

$$C_{12}H_{21}O_{11}$$

Calculated: m/z 341.2604
Found: 341.1078
Δ: 0.1526 amu

Figure 4.21

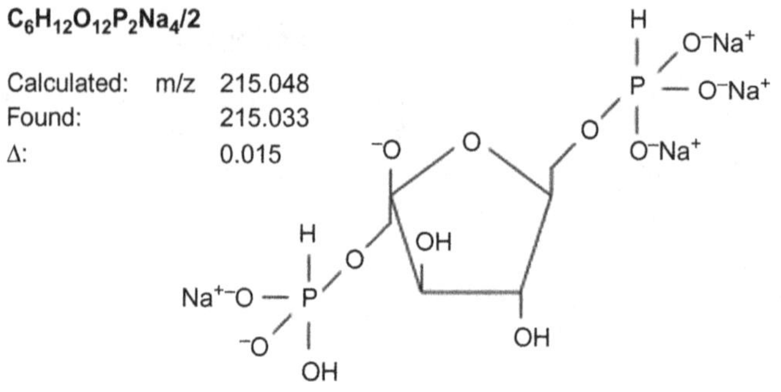

$C_6H_{12}O_{12}P_2Na_4/2$

Calculated: m/z 215.048
Found: 215.033
Δ: 0.015

Figure 4.22

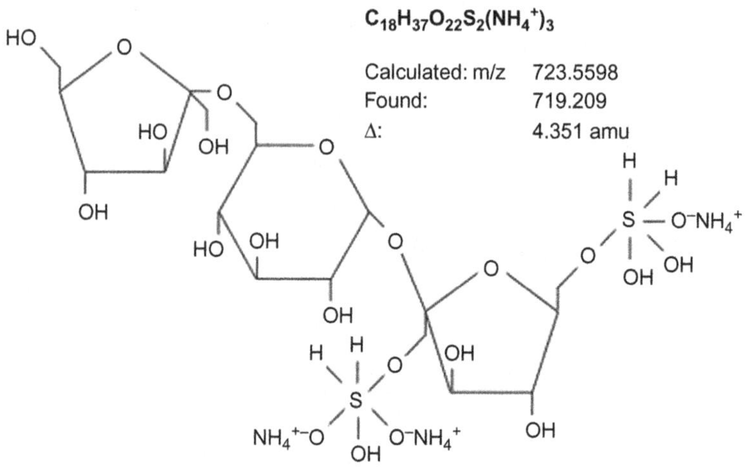

$C_{18}H_{37}O_{22}S_2(NH_4^+)_3$

Calculated: m/z 723.5598
Found: 719.209
Δ: 4.351 amu

Figure 4.23

The m/z 133 and m/z 115 ions' structures support the treatise that one hydride adds to the phosphate ester group. The daughter ion, m/z 115, loses only one ammonium ion to form the structure noted above from the structure for m/z 133. The m/z 115 ion has two negative charges. This is a common phenomenon for ions in MS/MS spectra and the reason why it occurs is unknown. The differences between the calculated and found ion masses are 0.07 for the m/z 133 and 0.03 for the m/z 115 ion. These differences are well within experimental error for this instrument and their small values lend credence to their identity as the structures drawn.

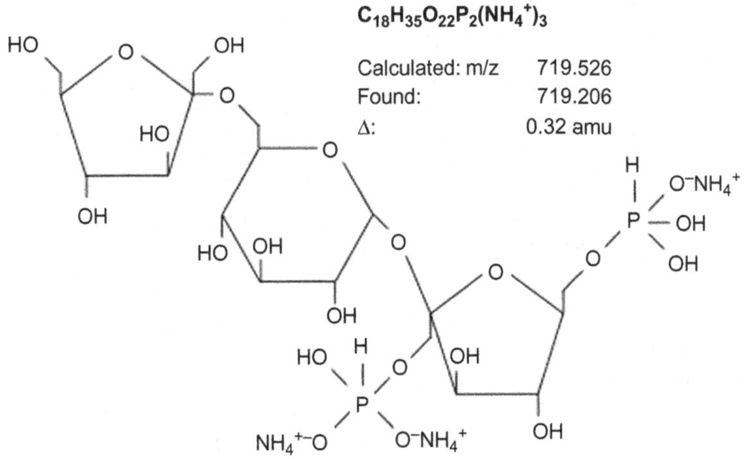

$$C_{18}H_{35}O_{22}P_2(NH_4^+)_3$$

Calculated: m/z   719.526
Found: 719.206
Δ: 0.32 amu

Figure 4.24

$$H_2PO_4(NH_4^+)_2$$

Calculated: m/z   133.0876
Found: 133.0145
Δ: 0.0731 amu

Figure 4.25

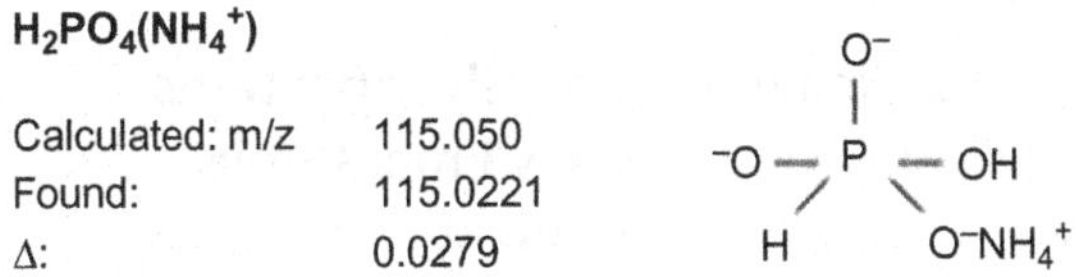

$$H_2PO_4(NH_4^+)$$

Calculated: m/z   115.050
Found: 115.0221
Δ: 0.0279

Figure 4.26

4.4 Method for Cleavage of Glycans From Glycoproteins

Because of the difficulties with other methods of glycan cleavage in both N-and O-linked oligosaccharides [1,2,32,38], we describe a novel method and carry out MS on one glycoprotein, Fetuin, for the N-linked oligosaccharides. We use this method to cleave, isolate and carry out MS of the O-linked oligosaccharides from the glycoproteins bovine sub-maxillary mucin, bovine thyroglobulin and K-casein (Fig. 4.27).

Protocol for the isolation of N- and O-linked oligosaccharide alditols is used to prepare N-linked oligosaccharide from fetuin and O-linked

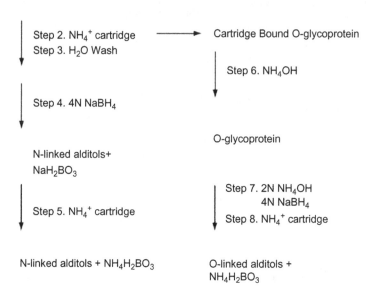

Step 1. Glycoprotein +PNGaseF

Step 2. NH$_4^+$ cartridge ———→ Cartridge Bound O-glycoprotein
Step 3. H$_2$O Wash

Step 6. NH$_4$OH

Step 4. 4N NaBH$_4$

O-glycoprotein

N-linked alditols+
NaH$_2$BO$_3$

Step 7. 2N NH$_4$OH
       4N NaBH$_4$
Step 5. NH$_4^+$ cartridge
Step 8. NH$_4^+$ cartridge

N-linked alditols + NH$_4$H$_2$BO$_3$ O-linked alditols +
                                              NH$_4$H$_2$BO$_3$

Figure 4.27

oligosaccharide alditols from bovine thyroglobulin, bovine submaxillary mucin and K-casein. In step 7 from the protocol above the molarity of NaBH$_4$ should be 12 μ equivalents or 3 μL of 4 N NaBH$_4$.

4.5 MS of Glycans Cleaved From Glycoproteins
4.5.1 MS of N-linked Glycan Alditols From Fetuin
From the following AQA single quadrapole mass spectrometer MS spectrum, we find evidence for tetra, tri-, and disialylated triantennary oligosaccharide linked to asparagine N-linkage. Note D = m/z 2880 and the base peak is D-3H^{-3} for the trisialylated triantennary N-linked alditols and the D-4H^{-4} for the tetrasialylated triantennary N-linked alditols. Please note the HPAEC-PAD for the mixture of N-linked oligosaccharide alditols was carried out on a CarboPac PA200 column. Harvey has published a similar spectrum for the N-linked oligosaccharide alditols from fetuin. Note also in the spectrum that there is little evidence of the presence of the disialylated triantennary oligosaccharide, possibly because of the milder method used to isolate these oligosaccharide alditols (Fig. 4.28).

MS of N-linked Oligosaccharide Alditols from Bovine Fetuin.

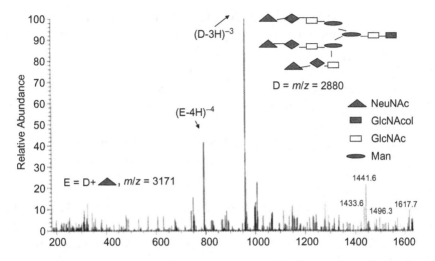

Figure 4.28 The negative ion mass spectrum, above, shows the N-linked oligosaccharide alditols from standard Fetuin, isolated in the manner described in the previous slide. It is quite clean and matches a standard spectrum found from Harvey's work.

4.6 Hydride Insertion Reduction Reaction Protocol

4.6.1 O-Linked Oligosaccharide Bound to Glycoprotein on Ammonium ion Cartridge

1. Elute from cartridge with 1 N NH₄OH. Immediately freeze. Vacuum dry until ~0.2 mL.
2. Make up to 1.0 mL with water.
3. Freeze.
4. Allow to thaw, partially.
5. Use 0.05 mL for reduction reaction.

4.6.2 O-Linked Oligosaccharide Bound to Glycoprotein

1. Make up to 1.0 mL with pH 11.4 ammonium hydroxide, immediately.
2. Add sodium borohydride (4 N, 3 μL).
3. Solution bubbles immediately.
4. Let stand at ambient temperature for 1 hour.
5. Immediately vacuum dry to ~0.1 mL.
6. Dissolve in 1.0 mL water.
7. Freeze. Partially thaw and analyze within 2 minutes by ESI negative ion triple quadrapole MS [25,26]

4.6.3 O-Linked Oligosaccharide Amino Acid Bound to Glycoprotein

1. Collect ammonium hydroxide eluent from ammonium form cation exchange resin.
2. Add sodium borohydride (4 N, 3.0 μL, 12 μeq).
3. Allow to stand capped at ambient temperature for 4 hours.
4. Uncap vial and allow to stand another 18 hours at ambient temperature.
5. Evaporate under vacuum at ambient temperature to ~0.2 mL.
6. Add 1.0 mL water to vial and freeze. Analyses were done on aliquots of partially frozen samples [20–23,36,39–42].

The chromatography of these O-linked oligosaccharide alditols/ O-linked oligosaccharide amino acids, noted below, on an MA1column with 310 mM NaOH as eluent attests to the purity of each alditol. They are shown in the next figure (Fig. 4.29A).

Note the relative purity of each oligosaccharide alditol mixture.

4.7 MS of Bovine Thyroglobulin O-Linked Oligosaccharide Alditol (Fig. 4.30)

Shown above is the MS from the reductive β-elimination of bovine thryoglobulin. Note the two major peaks, m/z 673.4 and m/z 265.3. They are structurally described in Fig. 4.31.

At a time later, we obtain the MS of O-linked [4–7] oligosaccharide alditol from bovine thyroglobulin. Note that the spectrum is similar to the previous spectrum, except that there is an m/z 674.4 and no m/z 265.3. (Fig. 4.32).

This ion, above, suggests that the glucose molecule is the reducing sugar in the original O-linked oligosaccharide glycoprotein. It agrees with the comparison between nonreductive β-elimination monosaccharide analysis, producing Glc and GlcNAc and Fucose. Whereas, the reductive β-elimination produces no Glc. The reducing sugar would be reduced and not appear as Glc, but glucitol, which is the above noted ion (Fig. 4.33).

Here is the cross-ring cleavage of the whole trisaccharide, which is desired over many cleavage products. This is done by using lower ionization temperatures in the API instrument (Fig. 4.34).

This ion, encompasses the trisaccharide in the reductive β-elimination reaction method (Fig. 4.35).

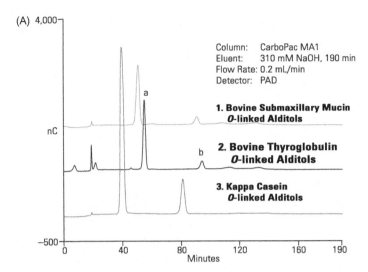

Figure 4.29A MALDI mixture of K-casein O-linked oligosaccharide alditols. In this MALDI MS we found an exact mass for the smaller, higher molecular weight peak as shown in the following figures. This is strong and direct evidence for the smaller of the two peaks being the 3', 6'-disialyl-4'-sulfo-4 hexosyl-6-sulfo-hexosaminitol, which possibly represents the smaller of the two chromatogram peaks found in the previous figure.

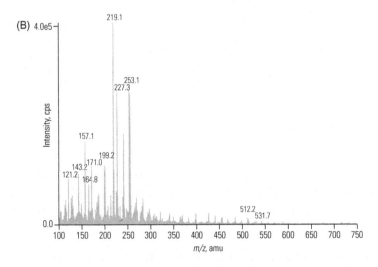

Figure 4.29B RBE of bovine submaxillary mucin O-linked oligosaccharides using excess ammonium hydroxide and minimal sodium borohydride.

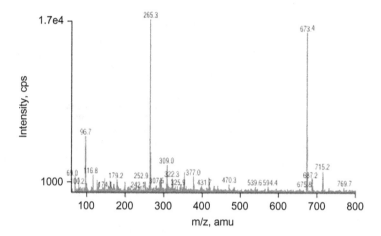

Figure 4.30

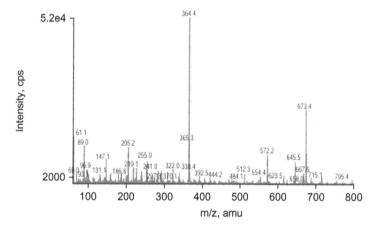

Figure 4.31

This is the MS/MS of spectrum of m/z 265 from bovine thyroglobulin oligosaccharide alditols. Note the indications of sulfation, m/z 97.2 and reduced sulfation of this ion. Note the two possible structures for m/z 97.2 in the next slide [4–6,25] (Fig. 4.36).

This ion has the noted possible structures and is from the MS/MS of m/z 265. This ion could come from two structures. Since this is reduced it is probably a dehydrated reduced sulfate bonded to a primary alcohol group. This would indicate sulfate substitution at either one or both primary positions from this molecule (A). The ion (B)

$C_6H_{17}O_9S$

Calculated: 265.2

Found: 265.3

Δ: 0.1 amu

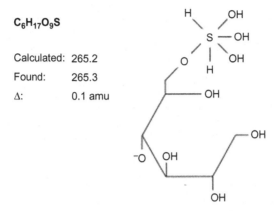

Figure 4.32

$(C_{17}H_{33}NO_{19}S_2(NH_4^+)_6)/2$

Calculated: 363.8
Found: 364.4
Δ: 0.6 amu

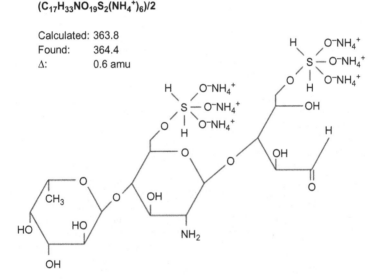

Figure 4.33

could be from the sulfate monoanion. For ion B to be found, cleavage would produce anionic oxygen, which, in water, would collapse to a double bond with concomitant loss of water using a proton from the solvent. This would produce uncharged sulfur trioxide gas as shown in the next slide (Fig. 4.37).

M/Z 79.8 comes from the MS/MS of m/z 265.0. We believe that the reduced sulfated carbon 6 to oxygen bond is cleaved. This ion then

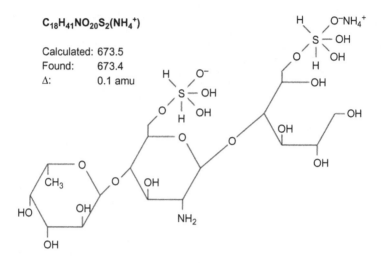

Figure 4.34

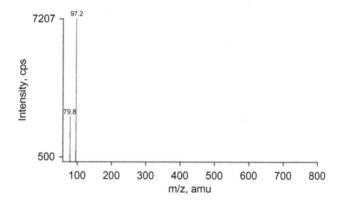

Figure 4.35

loses two molecules of H_2 gas, adds 1 proton, and loses 1 molecule of water to form the zwitterion of SO_3 gas. The evolving of H_2 and H_2O as gases serves to drive the reaction forward. Before the loss of water above, note the structure, negative oxygen may collapse to a double-bonded sulfur oxygen bond and produce m/z 97.2. This explains the other peak in the spectrum of MS/MS of m/z 265 (Fig. 4.38).

MS/MS of m/z peak 673 of bovine thyroglobulin O-linked alditols (Fig. 4.39).

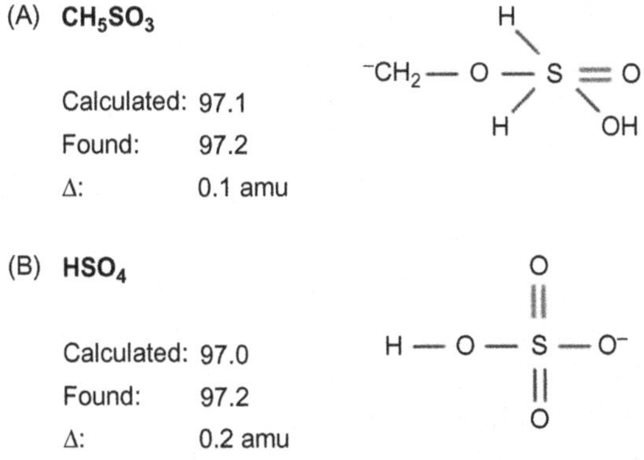

(A) CH5SO3
Calculated: 97.1
Found: 97.2
Δ: 0.1 amu
(B) HSO4
Calculated: 97.0
Found: 97.2
Δ: 0.2 amu

Figure 4.36

SO3(g)
Calculated: m/z 80.0
Found: 79.8
Δ: 0.2 amu

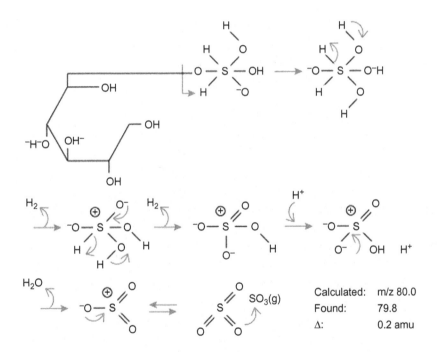

Figure 4.37

This ion is from the MS/MS of m/z of 673. It is of the glucosamine portion of the molecule. It is sulfated (reduced) and the acetamido group loses ketene, a common occurrence for acetamido sugars (Fig. 4.40).

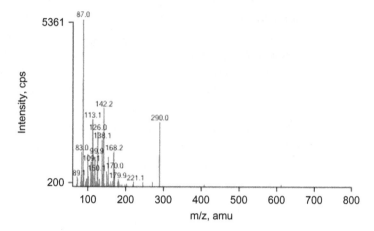

Figure 4.38

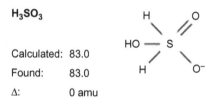

H₃SO₃

Calculated: 83.0
Found: 83.0
Δ: 0 amu

Figure 4.39

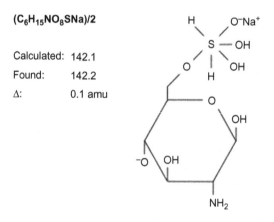

(C₆H₁₅NO₈SNa)/2

Calculated: 142.1
Found: 142.2
Δ: 0.1 amu

Figure 4.40

$C_6H_{13}NO_7SNa_2$

Calculated: 289.2
Found: 290.0
Δ: 0.8 amu

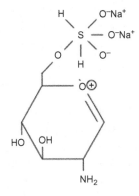

Figure 4.41

$(C_6H_{14}NO_8S)/3$

Calculated: 86.7
Found: 87.0
Δ: 0.3 amu

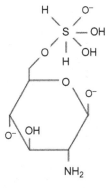

Figure 4.42

This ion, also from the MS/MS of m/z 673, is the reduced sulfate and dehydrated reduced sulfate (probably in the MS/MS sector). It is direct evidence for sulfation due to its reduction to the dihydrogen substitution [4–7,25]. The negative charge is possibly from cleavage between the carbon and oxygen of the reduced sulfate ester (Fig. 4.41).

This is an ion from the MS/MS of m/z 673. It is sulfated and reduced to an oxonium ion with a net negative charge. We have found that at the reducing end of the trisaccharide that an oxonium ion forms suggesting that, with the proposed cleavage at the four position, this is the penultimate monosaccharide in the trisaccharide. Sodiation, probably from glass walls of the spectrometer, happens in the MS/MS sector. It is derived from *N*-acetyl glucosamine (Fig. 4.42).

This is an ion from the MS/MS of m/z 673 again. It shows reduced sulfate at the primary position of glucosamine. There are two substitutions, one at the reducing end and another from the remaining part of the ion. This indicates that there is a glucosamine in the middle of the molecule and that it is sulfated at the primary position (Fig. 4.43).

Note the free amino group as in the other ions from m/z 673, which also have the free amino group. The reduced sulfate is dehydrated (probably in the mass spectrometer). Again there are two substitutions noted here at the 4′ position and the 6′ position (Fig. 4.44).

Here we see a possible reduced sulfated fucose ion, m/z 126.0. Rather than the second sulfation being on the 6 or 6′ primary groups, it is possibly on the 3″ position of fucose. It is also an oxonium ion (Fig. 4.45).

$(C_6H_{12}O_6NS)/2$

Calculated: 113.1

Found: 113.1

Δ: 0.0 amu

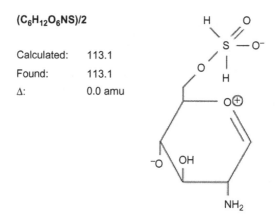

Figure 4.43

$(C_6H_{13}O_7SNa)/2$

Caluclated: 126.1

Found: 126.0

Δ: 0.1 amu

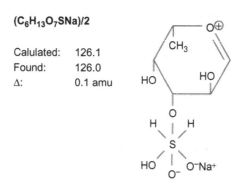

Figure 4.44

API-MS, on an API 2000 mass spectrometer, negative ion, of total sample of bovine thyroglobulin O-linked alditols, after reductively β-eliminated with ammonium hydroxide at pH 11.4 and sodium borohydride, was determined nearly immediately on preparation and thawing (Fig. 4.46).

This dianion is a fragment containing *N*-acetyl glucosamine and glucitol. Each of the monosaccharides having reduced sulfate at their

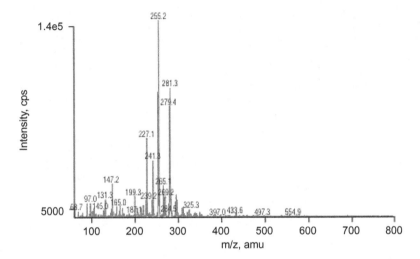

Figure 4.45

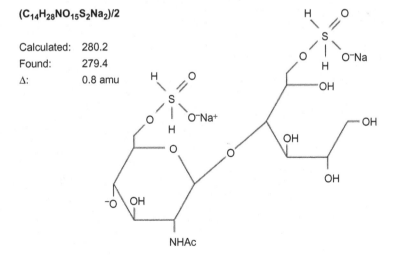

Figure 4.46

respective primary 6 positions. The sulfate is reduced and dehydrated with sodium ions as cations. This ion is also fragmented from the third monosaccharide, fucose (Fig. 4.47).

This structure is also a disaccharide dianion. It is substituted once with reduced sulfate and fragmented twice at the 6′ position and possibly the 4′ position. Glucitol is the "reducing end" alditol. This disaccharide alditol is composed of *N*-acetyl glucosamine and glucitol (Fig. 4.48).

(C$_{14}$H$_{27}$NO$_{14}$SNa$_2$)/2

Calculated: 255.7
Found: 255.1
Δ: 0.6 amu

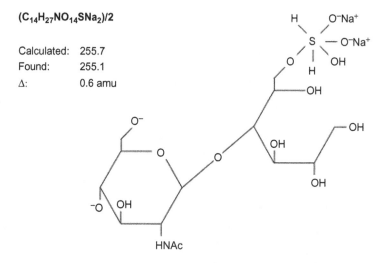

Figure 4.47

(C$_{14}$H$_{25}$NO$_{12}$SNa$^+$)/2

Calculated: 227.2
Found: 227.1
Δ: 0.1 amu

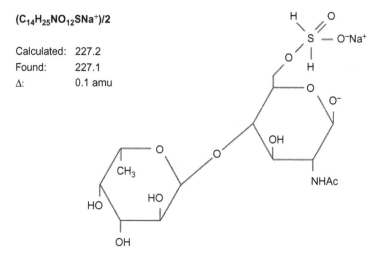

Figure 4.48

This dianion is composed of fucose and *N*-acetyl glucosamine. The reduced and dehydrated sulfate is shown on the 6 position of *N*-acetyl glucosamine. The GlcNAc is not reduced because it is not the reducing sugar in the complete molecule. This therefore supports the hypothesis that neither fucose nor *N*-acetyl glucosamine are reducing sugars of the complete molecule (Fig. 4.49).

This ion is derived from the fucose portion of the molecule, a cleavage product from the trisaccharide. It is cleaved only once which suggests that it is the nonreducing end of the molecule. Although it is difficult to envision how this negative ion came about, it is even more difficult to draw another structure for this ion (Fig. 4.49). The figure (Fig. 4.50) has 281.5 as its atomic mass unit. It is the full trisaccharide, containing a glucitol cross ring cleavage with glcNHAc ionized at the 6′ position. This is possibly where a di-hydrido sulfate group was previously found. It has an L-fucosyl group as the non-reducing end.

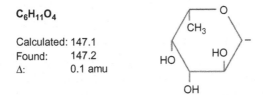

$C_6H_{11}O_4$

Calculated: 147.1
Found: 147.2
Δ: 0.1 amu

Figure 4.49

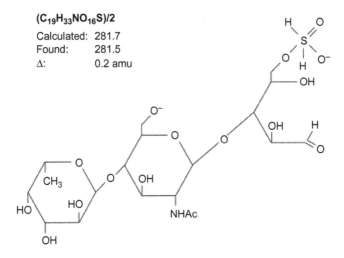

$(C_{19}H_{33}NO_{16}S)/2$

Calculated: 281.7
Found: 281.5
Δ: 0.2 amu

Figure 4.50

4.8 MALDI Mixture of K-Casein O-Linked Oligosaccharide Alditols/O-Linked Oligosaccharide Amino Acid

In this MALDI MS we found an exact mass for the m/z 1036.760 peak at higher amu than the largest peak height peak in this spectrum, m/z 698.857, as shown in the following figures (Fig. 4.51) [1,2,21,23,25–31]. This is strong and direct evidence for the smaller of the two peaks being the 3′, 6′-disialyl-4′-sulfo-4 hexosyl-6-sulfo-hexosaminitol, which possibly represents the smaller of the two chromatogram peaks found in the previous figure (Fig. 4.52).

This is the ion structure from the MALDI-TOF spectrum from the reductive β-elimination method with the O-linked oligosaccharide tyrosine from K-casein O-linked oligosaccharide (Fig. 4.53).

This is one of two ions possible for the smaller of the two peaks. The oligosaccharide is linked through an O-tyrosine glycosidic linkage (Fig. 4.54).

We propose the above structure for the main ion in the MALDI spectrum. It is only a trisaccharide and not sulfated. The difference between the calculated and found is 444 ppm. For MALDI of the second of the two peaks, the difference between the calculated and found is less than 26 ppm. This lack of precision for the larger peak may be because the MALDI is detecting two peaks, and the resolution is not adequate to compensate for the exchange rate of the protonated and unprotonated residues (Fig. 4.55).

Figure 4.51

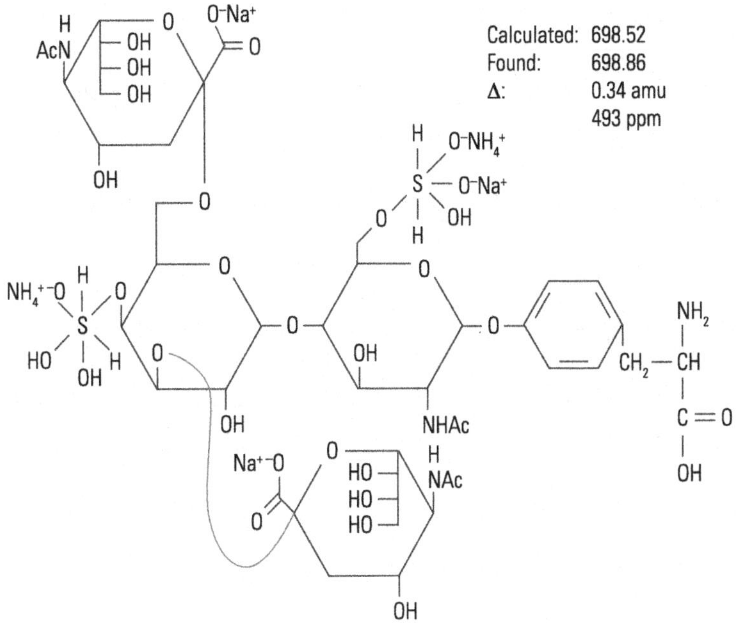

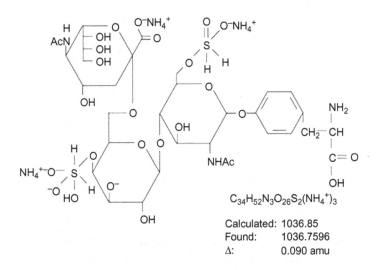

$(C_{45}H_{71}N_4O_{35}Na_3S_2(NH_4^+)_2)/2$

Calculated: 698.52
Found: 698.86
Δ: 0.34 amu
                      493 ppm

Figure 4.52

$C_{34}H_{52}N_3O_{26}S_2(NH_4^+)_3$

Calculated: 1036.85
Found: 1036.7596
Δ: 0.090 amu

Figure 4.53

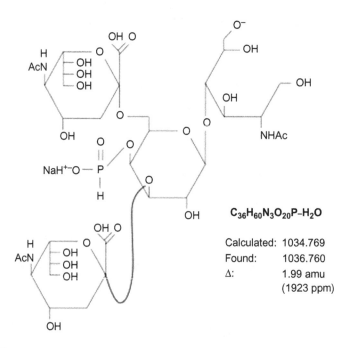

$C_{21}H_{43}O_{19}N_2Na$

Calculated: 698.549
Found: 698.8596
Δ: 0.31 amu

Figure 4.54

$C_{36}H_{60}N_3O_{20}P-H_2O$

Calculated: 1034.769
Found: 1036.760
Δ: 1.99 amu
                (1923 ppm)

Figure 4.55

We have drawn in this figure the possible ion structure that would be found if the molecule was phosphorylated. The difference between calculated and found ion mass is 1.99 amu or 1923 ppm, well outside the precision of the MALDI MS. Therefore this molecule is sulfated and not phosphorylated.

4.9 ESI-MS of K-Casein O-Linked Oligosaccharide Alditol/ O-Tyrosine Oligosaccharide (Fig. 4.56)

The spectrum above is from a single quadrapole ESI-MS of K-casein O-linked oligosaccharide alditol/O-tyrosine oligosaccharide. We note 5 ions (Fig. 4.57).

Here we have anion. possibly an O-tyrosine linked oligosaccharide. The difference between the calculated and found formula weights is 0.8 amu which is within the accuracy of the instrument. We would not be able to discern between sulfate versus phosphate ester substitution of doubly substituted doubly charged ion (Fig. 4.58).

This ion could be from either or both K-casein O-linked oligosaccharide alditol and the O-tyrosine oligosaccharide (Fig. 4.59).

This ion has the unsaturated serinyl tyrosine O-substituted structure, which could represent one of the ions from the single quadrapole ESI-MS (Fig. 4.60).

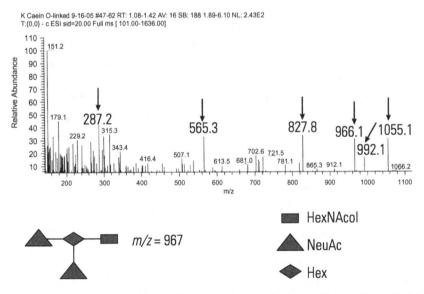

Figure 4.56 M/Z 966.1, Mi1H, di-sialyl hexosyl hexosaminitol, as noted by Saito and Itoh on a Dionex AQA single quadrapole mass spectrometer. It is reduced by sodium borohydride at pH 11.4 NH₄OH after β-elimination for the isolation of O-linked oligosaccharide alditols from O-linked glycoproteins. We will look at the five ions noted by arrows in the mass spectrum above. This spectrum is from the mixture of the two components of the O-linked oligosaccharide alditols shown by the chromatography noted. We find support for mono-dihydrido sulfate and for the di-dihydrido sulfate molecules which have been isolated from this mixture.

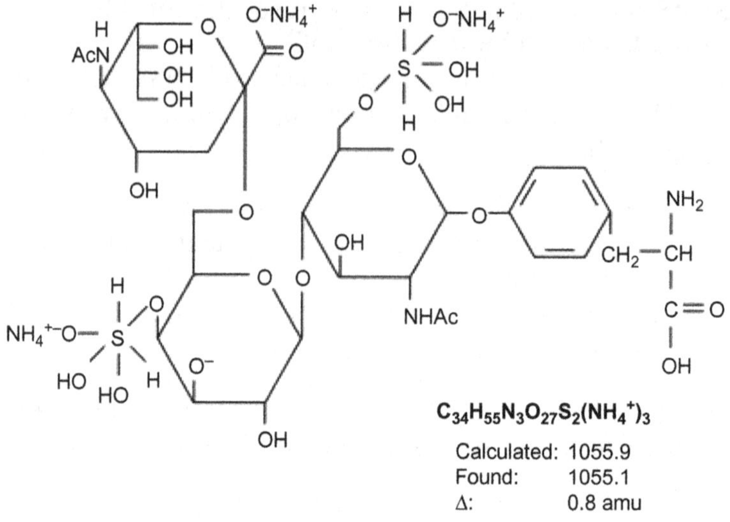

$$C_{34}H_{55}N_3O_{27}S_2(NH_4^+)_3$$

Calculated: 1055.9
Found: 1055.1
Δ: 0.8 amu

Figure 4.57

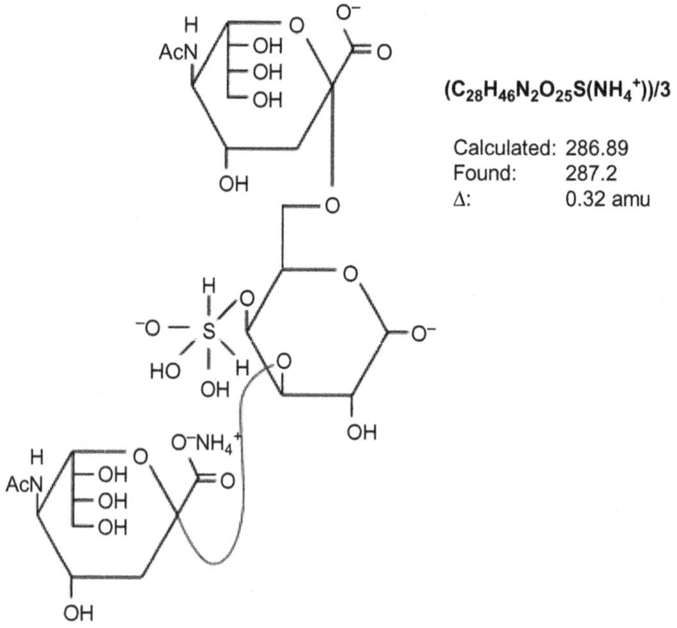

$$(C_{28}H_{46}N_2O_{25}S(NH_4^+))/3$$

Calculated: 286.89
Found: 287.2
Δ: 0.32 amu

Figure 4.58

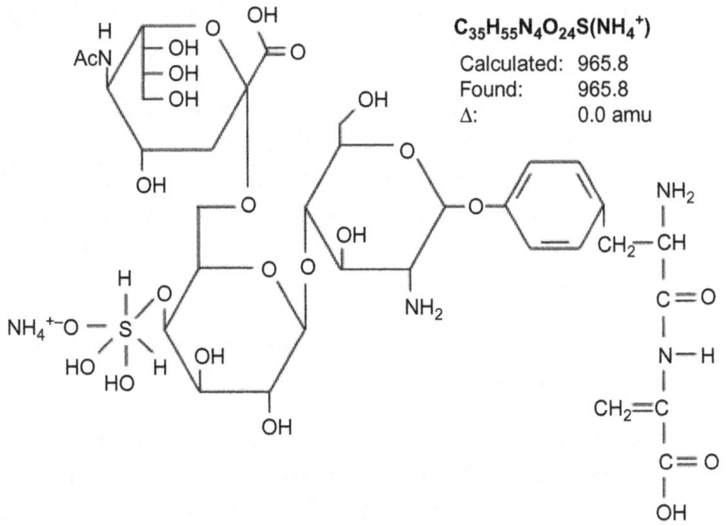

$C_{35}H_{55}N_4O_{24}S(NH_4^+)$

Calculated: 965.8
Found: 965.8
Δ: 0.0 amu

Figure 4.59

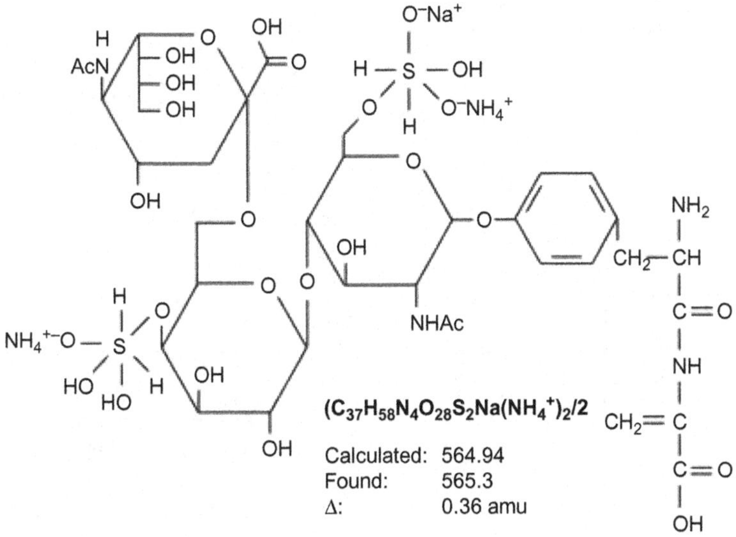

$(C_{37}H_{58}N_4O_{28}S_2Na(NH_4^+)_2/2$

Calculated: 564.94
Found: 565.3
Δ: 0.36 amu

Figure 4.60

Like the ion structure shown before this one, this could be a peptide substituted oligosaccharide from K-casein (Fig. 4.61).

This ion is another structural possibility for K-casein O-linked oligosaccharide alditols/O-linked oligosaccharide amino acid. This structure has two sulfo substitutions whereas the other one above indicates one substitution. This structure has four anionic oxygens counterbalanced with four sodium ions. This is possible because the walls of the AQA are glass and contain sodium ions (Fig. 4.62).

We have in this drawing a fragment that could be from the mono-sulfated molecule (Fig. 4.63).

This is a drawing of a mono-dihydrido sulfo compound. It is a singly cleaved structure at the anomeric oxygen of the disialyl hexosyl portion of the molecule. The original molecule could be either a mono- or di-substituted sulfated molecule (Fig. 4.64).

This figure shows the probable structure of the di-dihydrido sulfated disialylhexosylhexosaminitol. The difference between the calculated and found formula weight is 0.6 amu (Fig. 4.65).

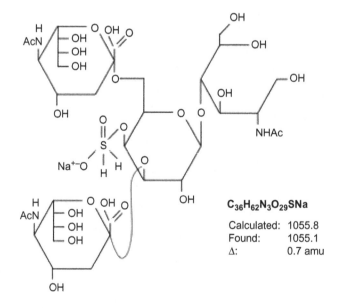

$C_{36}H_{62}N_3O_{29}SNa$

Calculated: 1055.8
Found: 1055.1
Δ: 0.7 amu

Figure 4.61

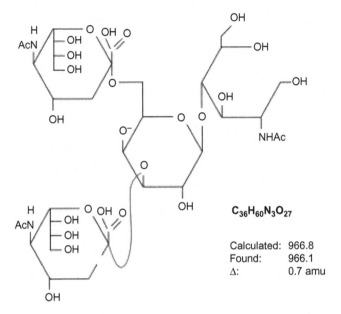

$C_{36}H_{60}N_3O_{27}$

Calculated: 966.8
Found: 966.1
Δ: 0.7 amu

Figure 4.62

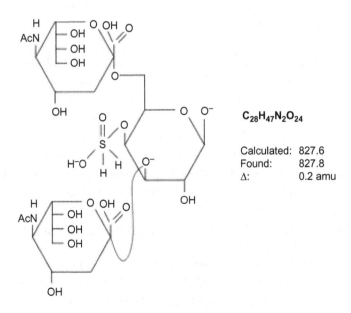

$C_{28}H_{47}N_2O_{24}$

Calculated: 827.6
Found: 827.8
Δ: 0.2 amu

Figure 4.63

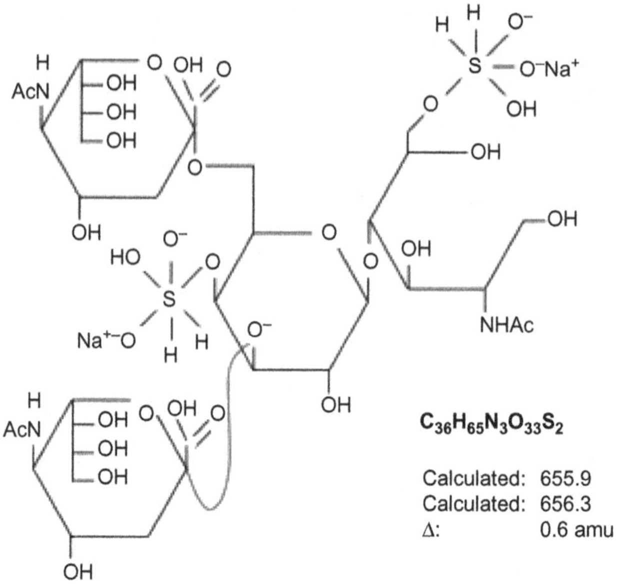

Figure 4.64

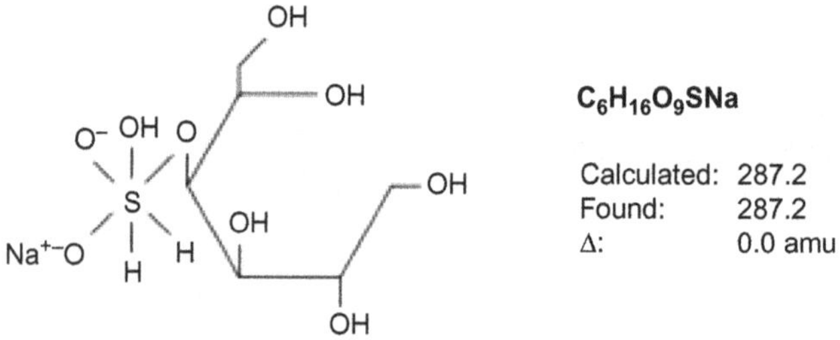

$C_{36}H_{65}N_3O_{33}S_2$

Calculated: 655.9
Calculated: 656.3
Δ: 0.6 amu

Figure 4.65

$C_6H_{16}O_9SNa$

Calculated: 287.2
Found: 287.2
Δ: 0.0 amu

This structure shows where one of the sulfo derivatives is located, on the hexosyl portion of the molecule. The difference between the calculated and found is 0 amu and indicates structural identity.

4.10 MS of O-Linked Oligosaccharide(s) of Bovine Submaxillary Mucin (Fig. 4.66)

Here we show the N-acetyl neuraminyl 1,4 disulfo-N-acetyl glucosamine and N-acetyl neuraminyl disaccharide. It has a calculated amu of 254.5 amu whereas the diphospho analog has a calculated amu of 253.1 amu. The difference between the calculated and found for

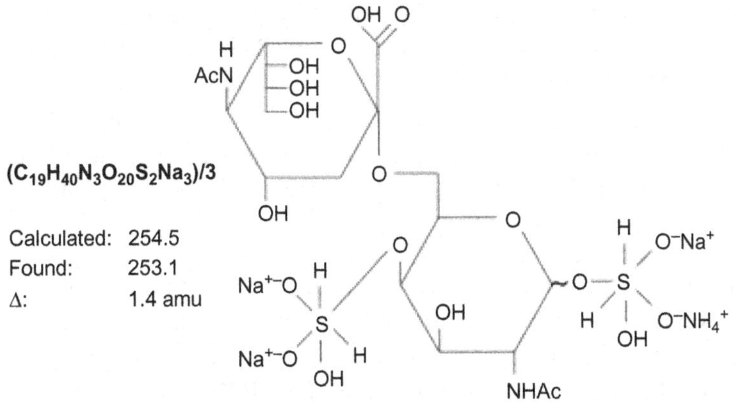
(C19H40N3O20S2Na3)/3
Calculated: 254.5
Found: 253.1
Δ: 1.4 amu
AcN
NHAc
O⁻Na⁺
O⁻NH4⁺

Figure 4.66

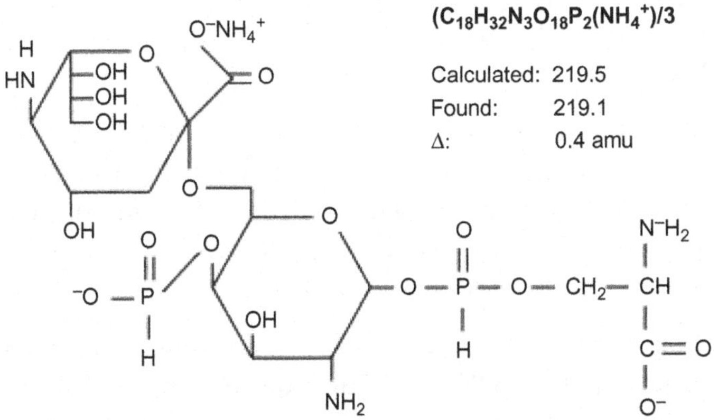
(C18H32N3O18P2(NH4⁺)/3
Calculated: 219.5
Found: 219.1
Δ: 0.4 amu
O⁻NH4⁺
HN
NH2
CH2
CH

Figure 4.67

the disulfo compound is 1.4 amu and for the diphospho ion is 0.0 amu. This suggests that the diphospho ion is the actual structure for this ion. The reason for the lowered difference between calculated and found is that it is a trianion with a charge to mass ratio of m/3. We probably cannot discern a monophospho versus a monosulfo carbohydrate ester thath as a mass to charge ratio of m/3 (Fig. 4.67).

In this figure we show a diphosphorylated *N*-acetyl neuraminyl glucosamine substituted serine. It has a 0.5 amu difference between the calculated and found chemical formula. This is a strong candidate for the actual structure of this anion. Note that the amino portion of glucosamine and neuraminic acid have no acetyl groups. This amide hydrolysis, for protein and peptides as well, is probably caused by a drop in pH to near pH 9–8. At this pH amides are hydrolyzed. Something similar to this happens when using hydrazine or hydrazine hydrate. The latter has a pK of around pH 8–8.5. This would explain at least the partial identity of the m/z 219.5 peak in the MS in Figs. 4.29A and 4.29B (Fig. 4.68).

Shown in Figs 4.68 and 4.69 is a proposed mechanism for the isolation of compound 3 and compound 4 in the product mixture using the reductive beta elimination (RBE) of bovine submaxillary mucin O-linked oligosaccharide glycoprotein. This RBE uses pH 11.4 ammonium hydroxide with sodium borohydride. It proposes that β-elimination under these circumstances is slower than amide hydrolysis of acetamido sugar and peptide and hydride insertion. We believe that amide hydrolysis is caused by the lowering of the reaction mixture pH to approximately pH [8–9], which is where amide hydrolysis occurs (Fig. 4.69).

This dianion is a cross-ring cleavage product of the whole molecule. It is cleaved between carbons 1 and 2 of the glucitol moiety. One of the 6 positions is substituted with dehydrated, reduced sulfate. The other 6 position hydroxyl has reduced sulfate cleaved from it. It is a clear indication of the trisaccharide structure (Fig. 4.70).

This is a MALDI spectrum of K-casein O-linked oligosaccharide alditol/amino acid [21,23,26–31]. Three out of the four ions are substituted by tyrosine, an amino acid. Since tyrosine-linked carbohydrates are not cleavable by β-elimination, they would persist. The remaining glycopeptide would be cleaved until the single amino acid remains (Fig. 4.71).

Here in this structure we have the whole molecule. It is the disodium, diammonium salt of the disulfo dihydrido hydrated sulfo derivative found in the second MALDI MS shown (Fig. 4.72).

Here we have the same molecule as in the previous structure except that there is the loss of one hydrogen ion and the addition of one sodium ion (Fig. 4.73).

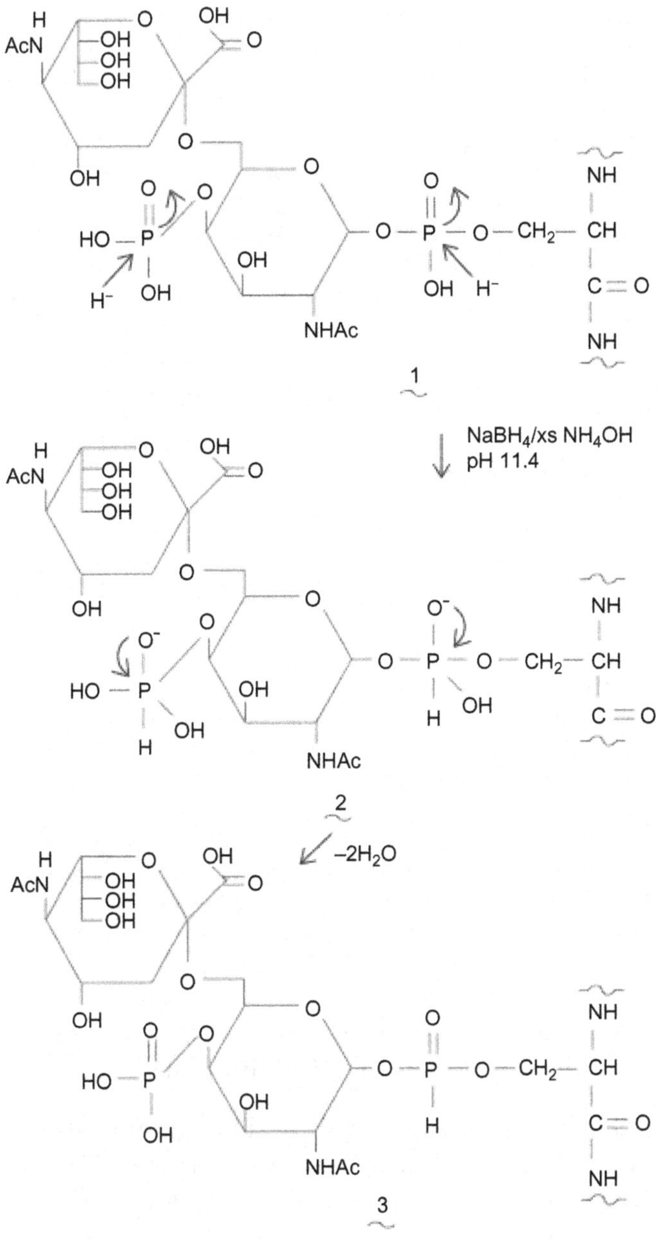

Figure 4.68

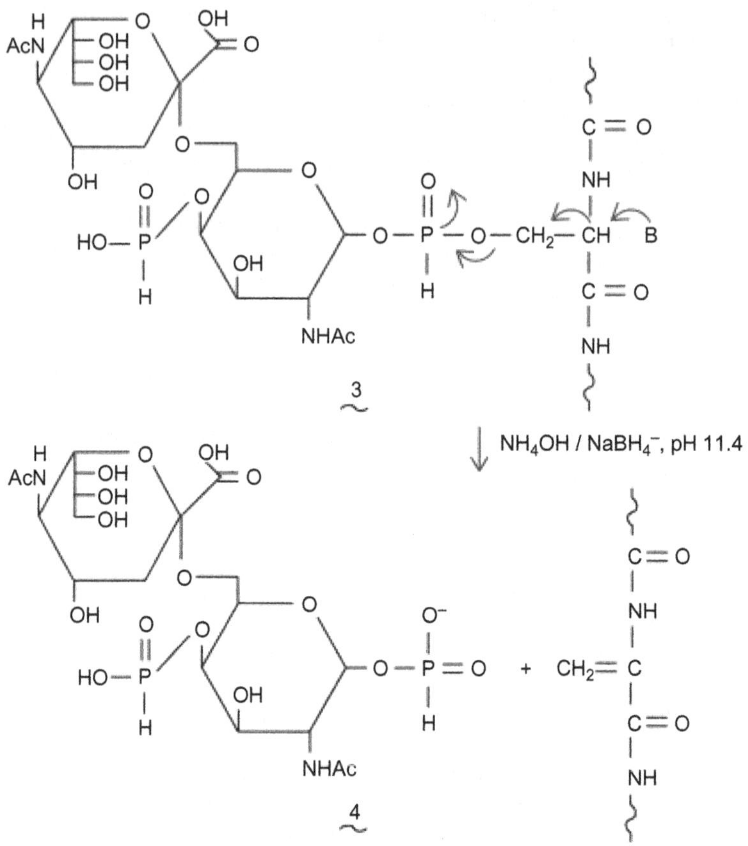

Figure 4.69

Here is another ion that has a formula from its structure that is within the accuracy of the MALDI-TOF instrument (Fig. 4.74).

The structure, above, is a dihydrido sulfo substituted derivative of 3′,6′ disialyl galactosyl galactosaminyl tyrosine [21,25]. It has completely deacetylated acetamido and peptido groups (Fig. 4.75).

This structure is from the MALDI MS from above. It does not show any tyrosine substitution. It is cleaved at the glycosidic linkage between galactose and galactosamine. It shows only one dihydrido sulfo group placed arbitrarily on O-4′ of the galactose residue [22,25,26,39–42].

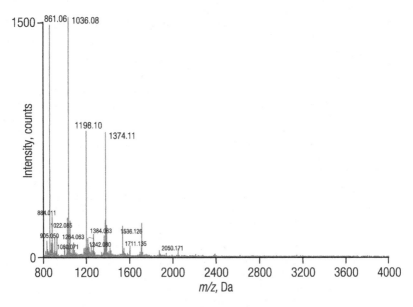

Figure 4.70 This Figure is a MALDI spectrum of K̲ casein O-linked oligosaccharide alditol/amino acid. Three out of the four ions are substituted by tyrosine, amino acid. Since tyrosine linked carbohydrates are not cleavable by β-elimination, they would persist. The remaining glycopeptide would be cleaved until the single amino acid remains.

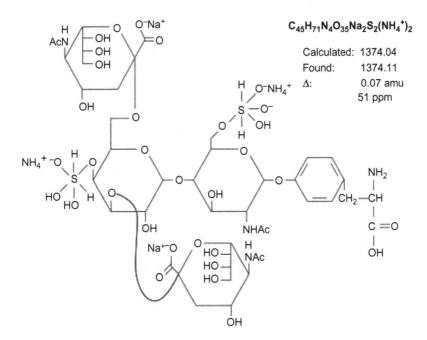

$C_{45}H_{71}N_4O_{35}Na_2S_2(NH_4^+)_2$

Calculated:	1374.04
Found:	1374.11
Δ:	0.07 amu
	51 ppm

Figure 4.71

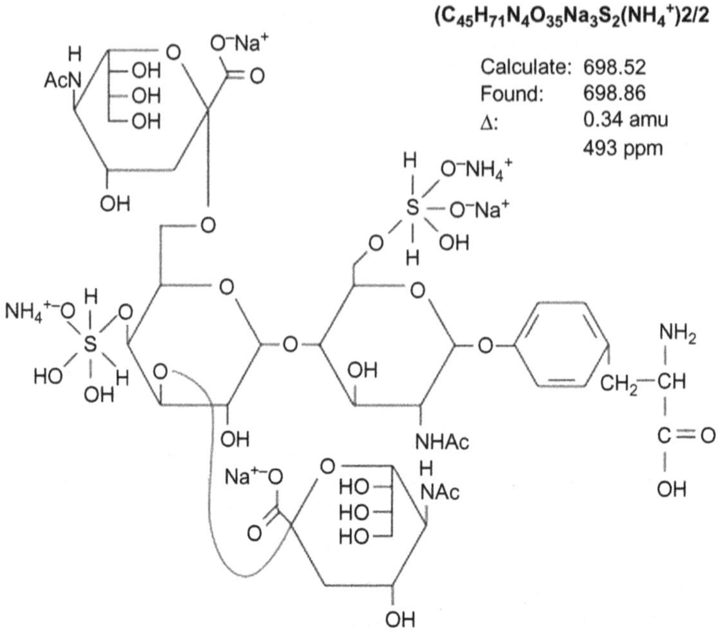

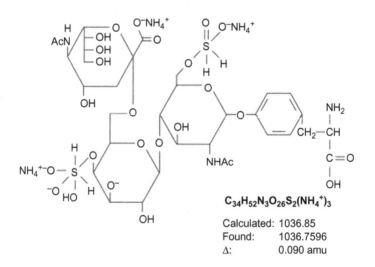

(C45H71N4O35Na3S2(NH4+)2/2
Calculate: 698.52
Found: 698.86
Δ: 0.34 amu
493 ppm
AcN
OH
O-Na+
O-NH4+
O-Na+
NH4+-O
HO
NHAc
NAc
Na+-O
HO
NH2
CH2—CH
C=O

Figure 4.72 This figure shows the structure found in the second MALDI-TOF of K casein O-linked oligosaccharide. It is tyrosine linked at the N-acetyl galactosaminyl O-1 position. The two ammonum ions are registered as part of the formula weight but the two charges are also seen by the MALDI-TOF MS.

AcN
OH
O-NH4+
NH4+-O—S
HO
NHAc
NH2
CH2—CH
C=O
C34H52N3O26S2(NH4+)3
Calculated: 1036.85
Found: 1036.7596
Δ: 0.090 amu

Figure 4.73

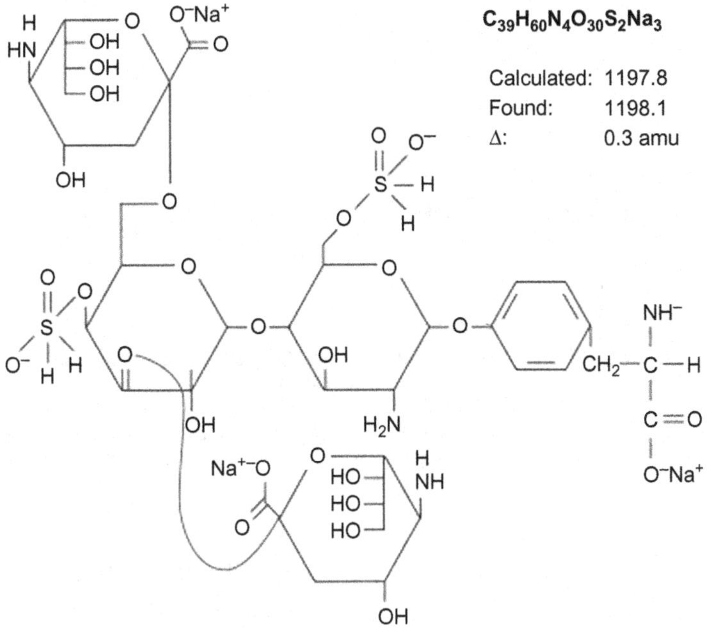

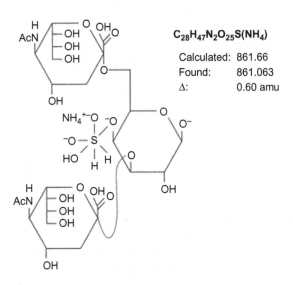

C39H60N4O30S2Na3

Calculated: 1197.8
Found: 1198.1
Δ: 0.3 amu

Figure 4.74

C28H47N2O25S(NH4)

Calculated: 861.66
Found: 861.063
Δ: 0.60 amu

Figure 4.75

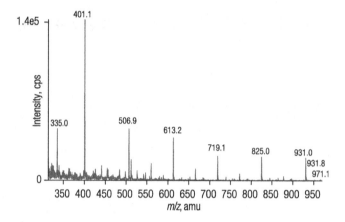

Figure 4.76 This spectrum was obtained by doing MS on the larger of the two peaks found in the MA1 chromatogram. This is relatively clean for the amount used to produce it. We describe the major ions in the figures that follow.

$(C_{46}H_{65}N_4O_{31}(NH_4^+)_2)/2$

Calculated: 613.0
Found: 613.2
Δ: 0.2 amu

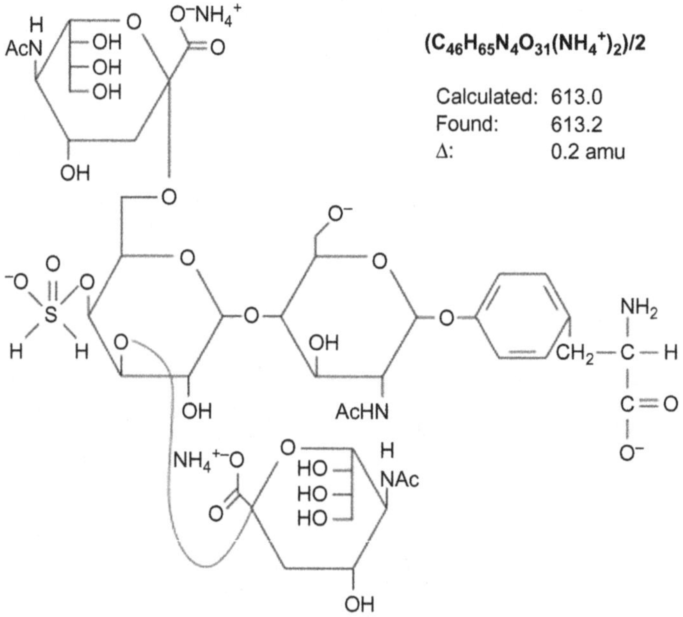

Figure 4.77

4.11 MS on the Larger of the Two Peaks Found in the MA1 Chromatogram (Fig. 4.76)

The spectrum above was obtained by carrying out MS on the larger of the two peaks found in the MA1 chromatogram. This is relatively clean for the amount used to produce it. We describe the major ions in the figures that follow (Fig. 4.77).

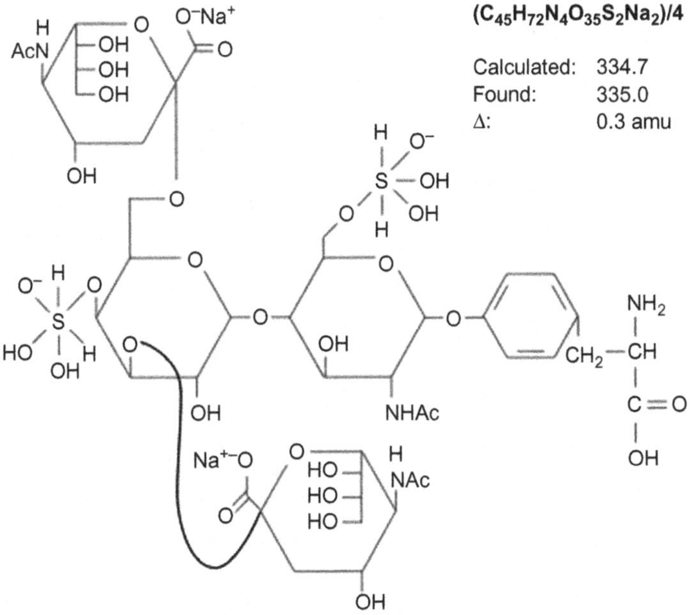

$(C_{45}H_{72}N_4O_{35}S_2Na_2)/4$

Calculated: 334.7
Found: 335.0
Δ: 0.3 amu

Figure 4.78

In Fig. 4.77 there is only one sulfo derivative with a cleavage where the second sulfo derivative could be found. The one sulfo derivative is dehydrated as well.

The structure drawn above is completely deamidated. This can be expected since there is only one amino acid substituted with an oligo-saccharide, the protein has been deamidated as well since peptides are actually polyamides (Fig. 4.78).

In the structure above, all monosaccharide residues have intact acetamido groups and fully hydrated dihydrido sulfo groups in the anionic form with no counter-balancing cation. The pH of amide hydrolysis for the acetamido groups must be higher than for peptido-amides (Fig. 4.79).

In this figure, *N*-acetyl neuraminic acid has been cleaved, sulfo groups are dihydrated and dehydrated with no loss of acetamido groups (Fig. 4.80).

This figure, which is similar to Fig. 4.38, shows that one of the dehydrated dihydrido sulfo groups has been hydrated and the acetamido groups and all of the peptide linkages hydrolyzed to the free amine (Fig. 4.81).

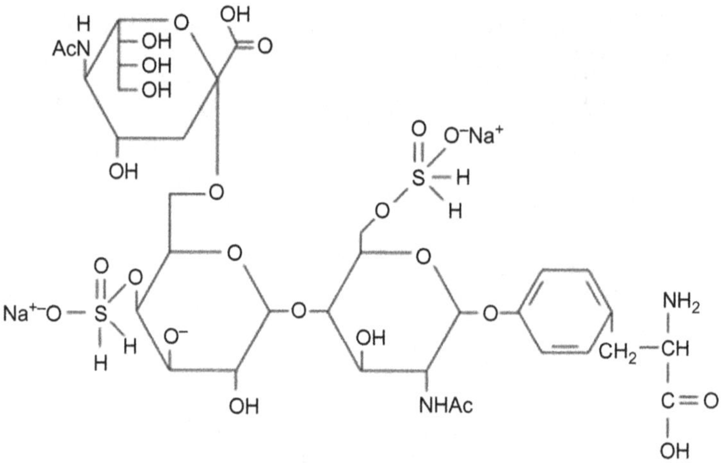

(C$_{34}$H$_{52}$N$_3$O$_{25}$S$_2$Na$_2$)/2

Calculated: 506.4
Found: 506.9
Δ: 0.5 amu

Figure 4.79

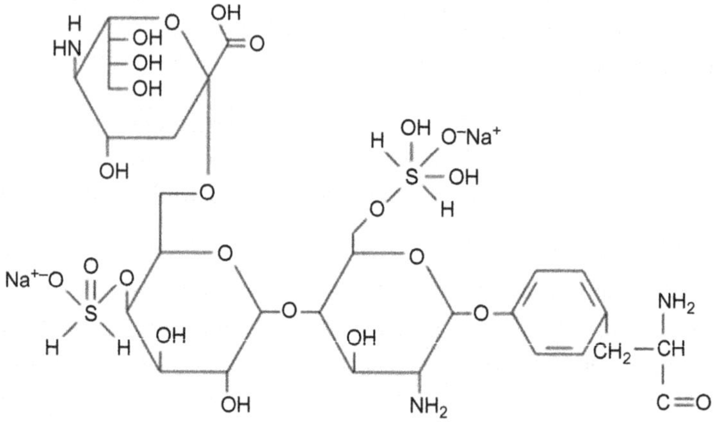

C$_{30}$H$_{50}$N$_3$O$_{33}$S$_2$Na$_2$

Calculated: 930.7
Found: 931.0
Δ: 0.3 amu

Figure 4.80

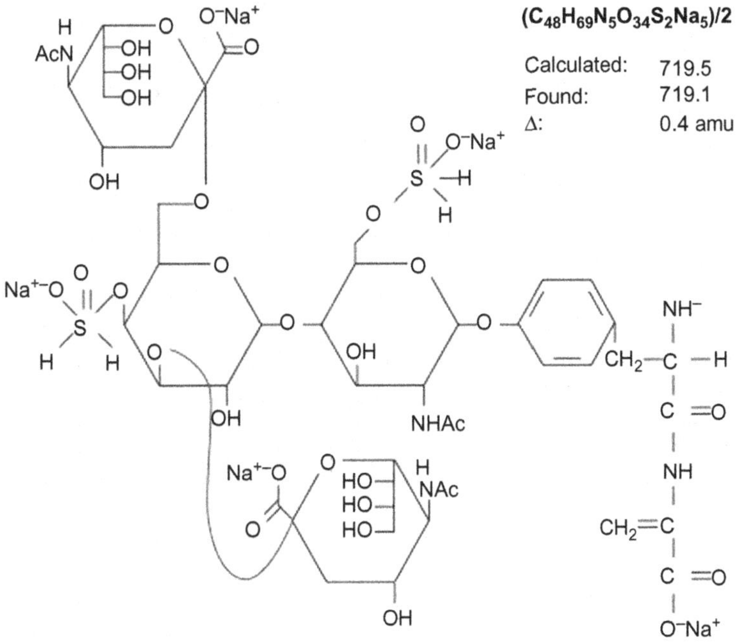

Figure 4.81

This figure, much like Fig. 4.37, has an additional amino acid linked to tyrosinyl oligosaccharide. It is the product of β-elimination of a carbohydrate linked to serine (Fig. 4.82).

This molecule is drawn above with cleavage between the galactosyl residue and N-acetyl galactosamine with charge retention at the galactosyl anomeric oxygen (Fig. 4.83).

This ion nearly demonstrates the identity of the original molecule. One ketene is lost from one of the sialic acid residues. One of the hexosyl residue oxygens is substituted with a hydride inserted dehydrated sulfate residue (Fig. 4.84).

This is a fragment ion that shows three substitutions on the hexosyl portion of the whole molecule. This ion is the whole molecule fragmented at O-1′ with substitution at O-6′ and O-3′, sialic acid residues, and a hydride inserted sulfate residue at O-4′ of the hexosyl residue. The difference between calculated and found amu is 0.3 amu, suggesting the identity of this ion as the structure shown (Fig. 4.85).

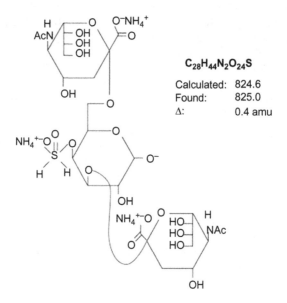

$C_{28}H_{44}N_2O_{24}S$

Calculated: 824.6
Found: 825.0
Δ: 0.4 amu

Figure 4.82

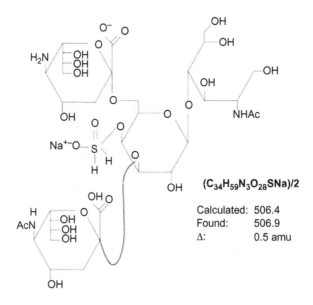

$(C_{34}H_{59}N_3O_{28}SNa)/2$

Calculated: 506.4
Found: 506.9
Δ: 0.5 amu

Figure 4.83

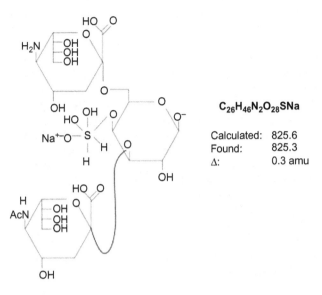

$C_{26}H_{46}N_2O_{28}SNa$

Calculated: 825.6
Found: 825.3
Δ: 0.3 amu

Figure 4.84

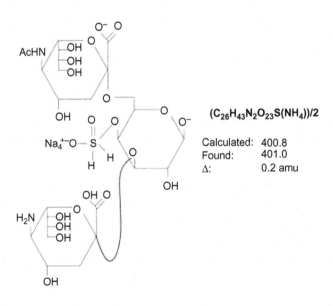

$(C_{26}H_{43}N_2O_{23}S(NH_4))/2$

Calculated: 400.8
Found: 401.0
Δ: 0.2 amu

Figure 4.85

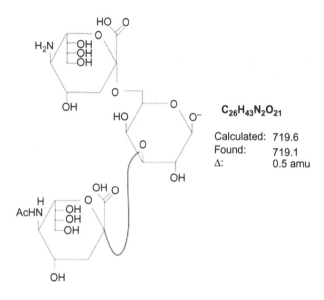

$C_{26}H_{43}N_2O_{21}$

Calculated: 719.6
Found: 719.1
Δ: 0.5 amu

Figure 4.86

This is a fragment of the original molecule, the larger peak of the two peaks isolated. It contains two sialic acid residues substituting the hexosyl group. The only difference between this structure and the previous one, except for ammonium versus sodium in the previous slide, is that this one is dehydrated. One of the sialic acid groups has lost ketene. This is the base peak with a 0.2 amu difference between the calculated and found amu presumably indicating identity (Fig. 4.86).

This trisaccharide fragment is cleaved from the original molecule at the anomeric oxygen with two substitutions, both sialic acid derivatives. One of the acetamido groups loses ketene and produces a free amino group (Fig. 4.87).

Here we find the whole molecule fragmented at the O-4′ position of the hexosyl residue less two molecules of water. We believe this O-4′ substitution is where the sulfate is in the original molecule. The difference between the calculated and found ion mass amu is 0.2 amu, well within the precision of the API 2000 Mass Spectrometer (Fig. 4.88).

Here we have a portion of the original ion that presumably had two substitutions hydrolyzed before the magnetic sector. We believe that the sulfate derivative and one sialic acid were hydrolyzed. Because the

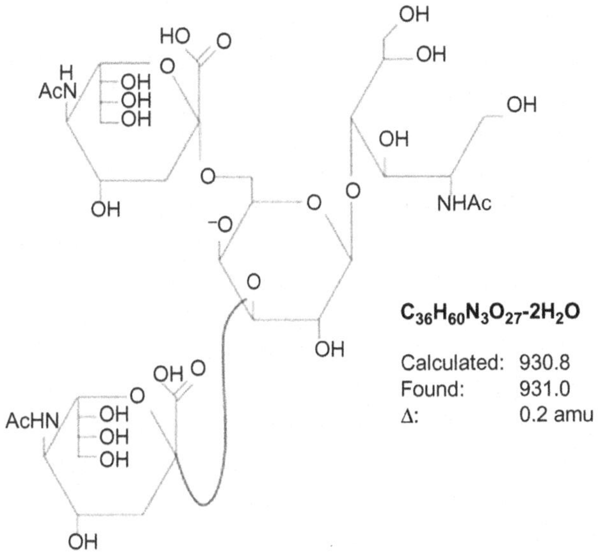

$C_{36}H_{60}N_3O_{27}\text{-}2H_2O$

Calculated: 930.8
Found: 931.0
Δ: 0.2 amu

Figure 4.87

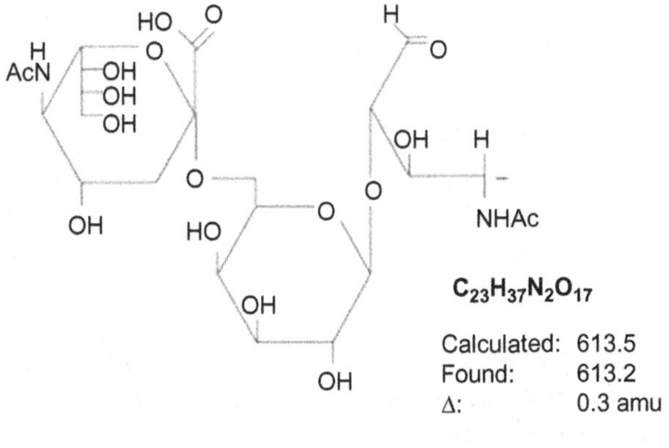

$C_{23}H_{37}N_2O_{17}$

Calculated: 613.5
Found: 613.2
Δ: 0.3 amu

Figure 4.88

difference between calculated and found ion mass is 0.3 amu, we believe this structure to be correct (Fig. 4.89).

This ion, from the larger of the two peaks in the MA1 chromatogram, is the product of two cross-ring cleavages, one between C-5 and C-6, and the other between C-2 and O-5. This ion is evidence for the original monosulfated tetrasaccharide from K-casein.

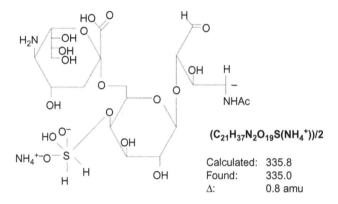

$(C_{21}H_{37}N_2O_{19}S(NH_4^+))/2$

Calculated: 335.8
Found: 335.0
Δ: 0.8 amu

Figure 4.89

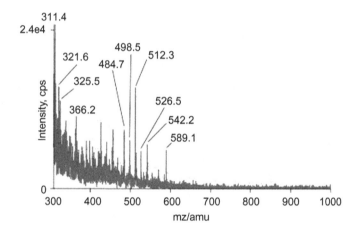

Figure 4.90

4.12 MS Generated From Isolating the Smaller of Two Peaks From the MA1 Chromatogram of K-Casein O-Linked Oligosaccharide Alditols (Fig. 4.90)

This is the mass spectrum generated from isolating the smaller of two peaks from the MA1 chromatogram of K-casein O-linked oligosaccharide alditols (Fig. 4.91).

Here is evidence for disulfation of the smaller peak. It is the "base peak" of the spectrum. This is a tetra ammonium salt of hydride inserted sulfate with four ammonium salts. The difference between the calculated and found ion mass, 0.6 amu, is well within the instrument's precision (Fig. 4.92).

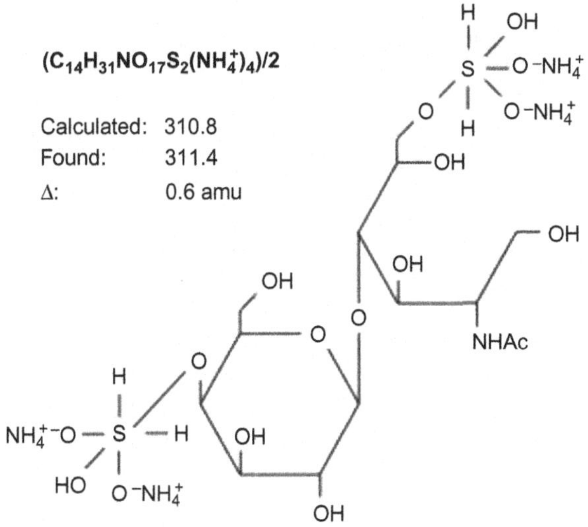

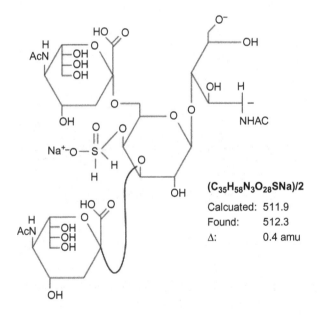

$(C_{14}H_{31}NO_{17}S_2(NH_4^+)_4)/2$
Calculated: 310.8
Found: 311.4
Δ: 0.6 amu
OH
O-NH4+
NHAc

Figure 4.91

HO
AcN
OH
NHAC
Na+-O
$(C_{35}H_{58}N_3O_{28}SNa)/2$
Calcuated: 511.9
Found: 512.3
Δ: 0.4 amu

Figure 4.92

This dianion shows all substitutions of the whole molecule, two sialyl residues and two sulfated residues, one outright, and the other deduced from an ion cleavage. The difference between the calculated and found mass is 0.4 amu, well within the precision of the API 2000

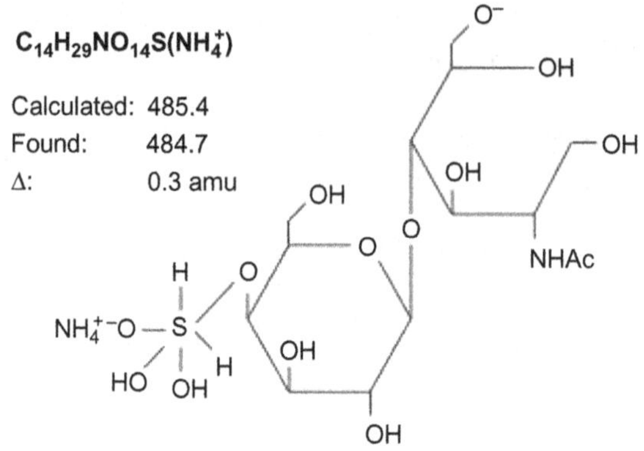

$C_{14}H_{29}NO_{14}S(NH_4^+)$

Calculated: 485.4
Found: 484.7
Δ: 0.3 amu

Figure 4.93

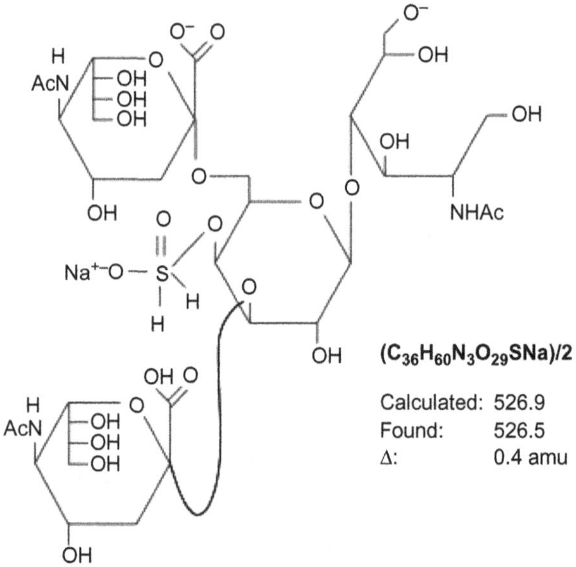

$(C_{36}H_{60}N_3O_{29}SNa)/2$

Calculated: 526.9
Found: 526.5
Δ: 0.4 amu

Figure 4.94

mass spectrometer. This ion is a cross-ring cleavage between C-1 and O-5 of the hexosaminitol (Fig. 4.93).

In this figure we have the neutral disaccharide with two substitutions shown: one at O-6, presumed sulfate substitution, one at O-4′, actual hydride inserted sulfate indicating underivatized sulfate in the original molecule. The difference between the calculated and found ion mass is 0.3 amu, well within the API 2000's precision (Fig. 4.94).

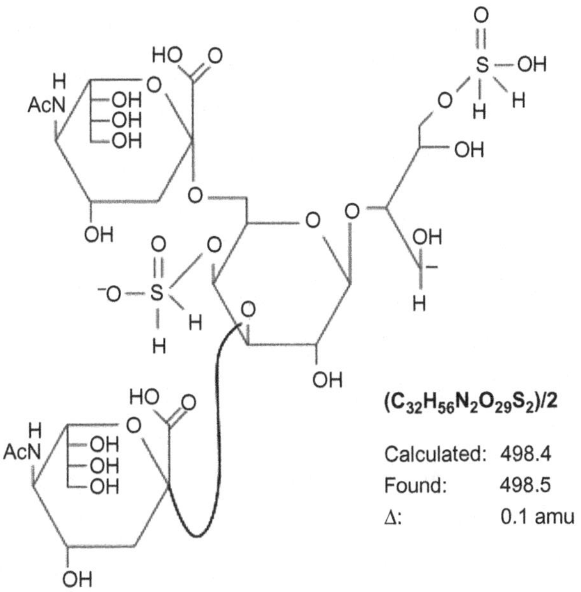

$(C_{32}H_{56}N_2O_{29}S_2)/2$

Calculated: 498.4
Found: 498.5
Δ: 0.1 amu

Figure 4.95

In this slide we have substitution by hydride inserted and dehydrated sulfate at the hexosyl residue. We also have cleavage at the 6 position of the *N*-acetyl hexosaminitol residue. This ion's structure supports the hypothesis of a disulfated molecule for the smaller of the two chromatographed O-linked oligosaccharide alditols from K-casein (Fig. 4.95).

Here, another cross-ring cleavage between O-5 and C-3, is an ion that directly represents the four substitutions of the original molecule from the O-linked oligosaccharide of K-casein. The difference between the calculated and found is 0.1 amu, suggesting that this is the structure of the ion (Fig. 4.96).

This is an ion fragment that shows three possible substitutions: one at O-6′ of the sialyl group, one at the O-4′ of the hydride inserted and reduced sulfate group, and one unknown substitution at O-3′ of the hexosyl residues (Fig. 4.97).

Again we have all four substitutions of the hexosyl *N*-acetyl hexosaminitol shown at least by fragmentations or directly. The difference between calculated and found amu is only 0.1 amu. Which suggests that the structure above is the actual ion's structure (Fig. 4.98).

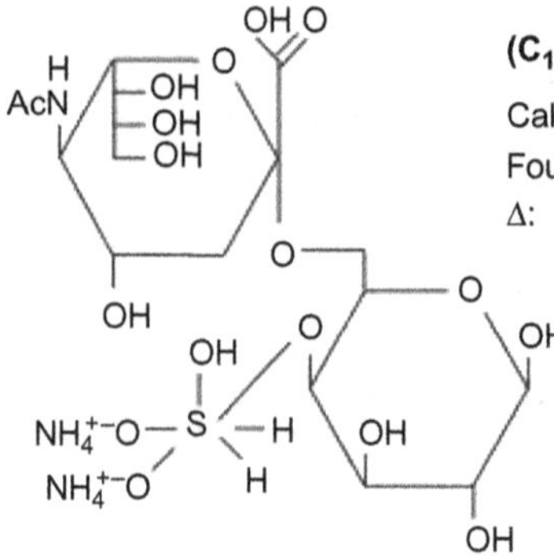

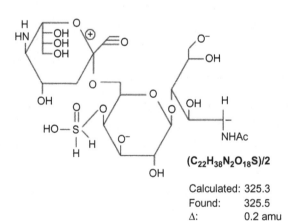

$(C_{17}H_{31}NO_{17}S(NH_4^+)_2)/2$

Calculated: 589.5
Found: 589.1
Δ: 0.4 amu

Figure 4.96

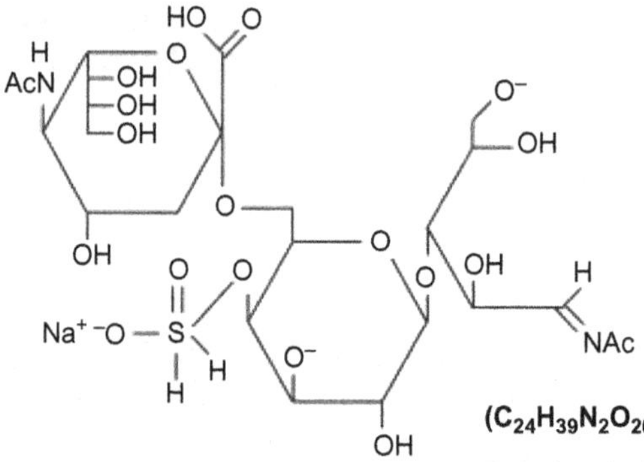

$(C_{24}H_{39}N_2O_{20}SNa)/2$

Calculated: 365.3
Found: 365.2
Δ: 0.1 amu

Figure 4.97

$(C_{22}H_{38}N_2O_{18}S)/2$

Calculated: 325.3
Found: 325.5
Δ: 0.2 amu

Figure 4.98

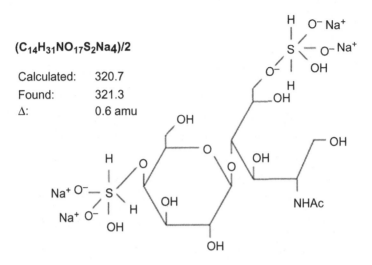

Figure 4.99

We have here a carboxylium ion of sialic acid, substituted on the hexosyl structure at the 6' position. There is dehydrated, hydride inserted sulfate at the 4' position and an unknown substitution at 3' of the hexosyl group, possible sialic acid as substitution, and substitution at the 6 position, possibly a dihydrido sulfo substitution (Fig. 4.99).

This structure proposes both the hydride inserted and the reduced hexosaminitol and is evidence that the smaller of the two peaks is disulfated (Fig. 4.100).

In this figure we show a structure that has 4 substitutions of Hex-HexNAc. There are to sialyl groups, one hydride inserted, dehydrated sulfate ester and one substitution that is cleaved to give the molecule net one negative charge. We believe this is the second hydride, dehydrated substituted sulfate ester. The difference between calculated ion mass and found ion mass is less than 26 ppm. This can then be considered exact mass for this ion and its structural identity (Fig. 4.101).

We propose the above structure for the main ion in the MALDI spectrum. It is only a trisaccharide and not sulfated. The difference between the calculated and found is 444 ppm. For MALDI of the second of the two peaks, the difference between the calculated and found

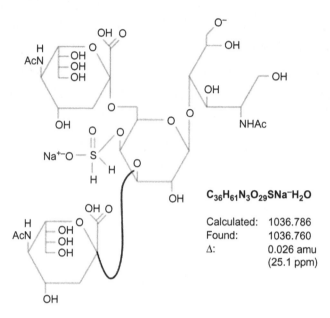

$C_{36}H_{61}N_3O_{29}SNa^-H_2O$

Calculated: 1036.786
Found: 1036.760
Δ: 0.026 amu
             (25.1 ppm)

Figure 4.100

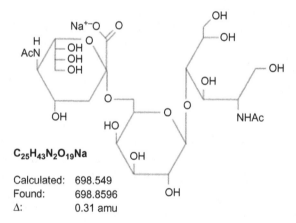

$C_{25}H_{43}N_2O_{19}Na$

Calculated: 698.549
Found: 698.8596
Δ: 0.31 amu

Figure 4.101

is less than 26 ppm. This lack of precision for the larger peak may be because the MALDI is detecting two peaks, and the resolution is not adequate to compensate for the exchange rate of the protonated and unprotonated residues (Fig. 4.102).

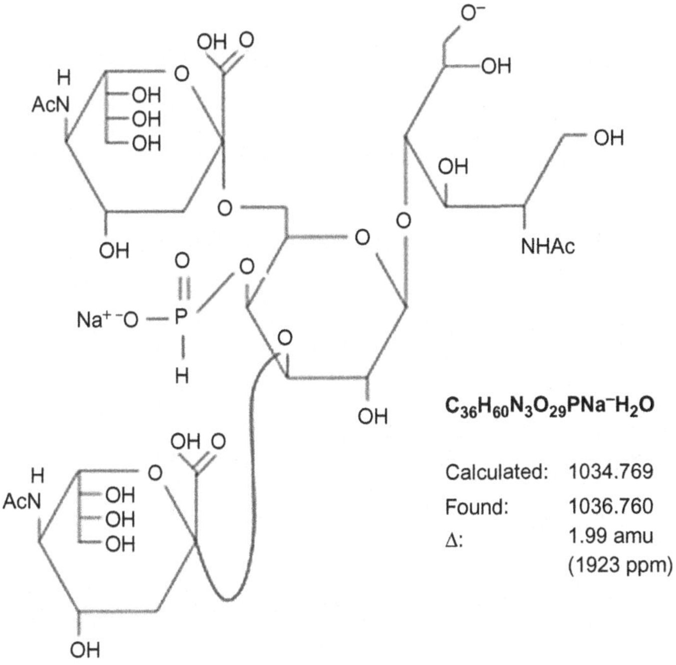

C_{36}H_{60}N_3O_{29}PNa^-H_2O

Calculated: 1034.769
Found: 1036.760
Δ: 1.99 amu
              (1923 ppm)

Figure 4.102

We have drawn in this figure the possible ion structure that would be found if the molecule was phosphorylated. The difference between the calculated and found ion mass is 1.99 amu or 1923 ppm, well outside the precision of the MALDI MS. Therefore this molecule is sulfated and not phosphorylated.

4.13 Possible Mechanism as to How Both K-Casein O-Linked Oligosaccharide-Tyrosine and O-Linked Oligosaccharide Alditol Are Produced (Fig. 4.103)

Here, above is a possible mechanism as to how both K-casein O-linked oligosaccharide-tyrosine and O-linked oligosaccharide alditol could occur in the same sample [21,23,26–31]. If the first sugar linked to peptide is a 2-acetamido-2-deoxy hexosyl group then it can form an oxazoline ring structure involving sugar carbon 1, carbon 2 and the acetamido group. This oxazoline can form at pHs above the pKa of the amido-proton, ~9.5. At pHs above 9.5 this proton can be removed to form a C N double bond, which not only prevents amide hydrolysis but also puts a good nucleophile(−O^−) in the vicinity of the anomeric

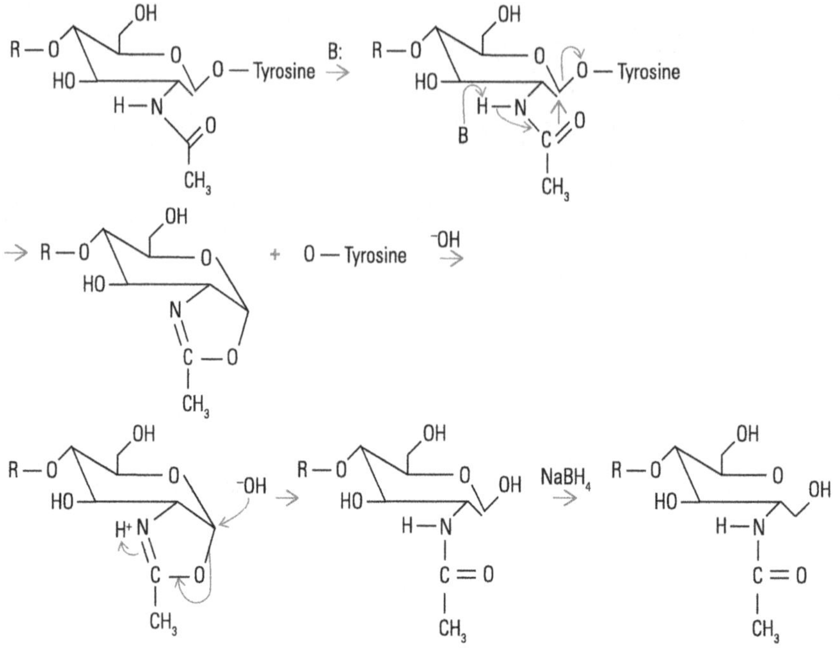

Figure 4.103

carbon. The bonding of this oxygen atom to the anomeric carbon kicks out ‾O-tyrosine. The hemiacetal is then reduced to the alditol by sodium borohydride.

Because the original method for the cleavage and isolation of N- and O-linked oligosaccharides [1,2,38] converted sulfated and phosphorylated oligosaccharides to their dihydrido and mono-hydrido derivatives, we looked for and discovered a method for the nonreductive β-elimination of O-linked oligosaccharides. We have done the MS of the O-linked oligosaccharide from bovine thyroglobulin glycoprotein. The following section presents the MS and the interpretation of the MS. But first we discuss the HPAEC-PAD of the component monosaccharides after hydrolysis by HCl (at low temperature and higher temperature).

4.14 Standard Monosaccharides Alone With Hydrolyzed nonreductively β-Eliminated O-Linked Oligosaccharides From Bovine Thyroglobulin Spiked With Standards (Fig. 4.104)

We compare standard monosaccharides alone with hydrolyzed nonreductively β-eliminated O-linked oligosaccharides from bovine

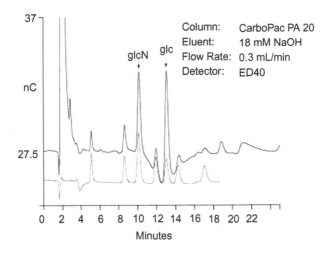

Figure 4.104 Base catalyzed hydrolysis and reduction of 2 deoxyglyco-O-tyrosine linkage.

thyroglobulin spiked with standards. We find GlcNAc and Glc to be component sugars of this oligosaccharide from bovine thyroglobulin. The monosaccharide analysis with reductive β-elimination gives GlcNAc, suggesting that glucose is the reducing sugar [7].

4.15 Mild Acid Hydrolysis of Bovine Thyroglobulin O-Linked Alditols (Fig. 4.105)

Mild acid hydrolysis of bovine thyroglobulin O-linked alditols is designed to release fucose and other acid sensitive monosaccharides. Note the peak matching fucose retention on the CarboPac PA 20 column. The upper trace is a mix of six standard monosaccharides. This figure demonstrates that fucose is a component monosaccharide of bovine thyroglobulin O-linked oligosaccharide.

4.16 MS, Negative Ion, of Nonreductive β-Elimination of Bovine Thyroglobulin O-Linked Oligosaccharide on API 2000 (Fig. 4.106)

The figure above shows API MS, negative ion, of nonreductive β-elimination of bovine thyroglobulin O-linked oligosaccharide on API 2000 (Fig. 4.107) [4–7].

This ion, m/z 89.2, is also one of the major ions found in the non-reductive beta elimination mass spectrometry (NRBE MS). It shows two substitutions on an unreduced molecule of glucose. Here is

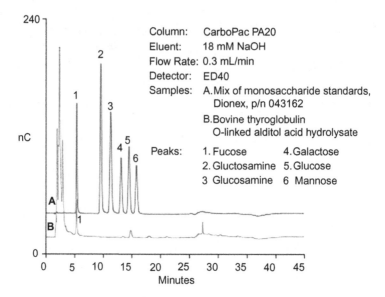

Figure 4.105

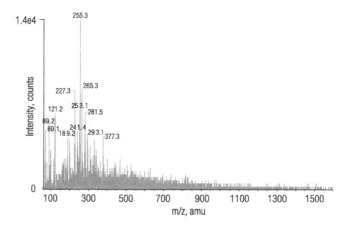

Figure 4.106

evidence that this oligosaccharide is the reducing sugar and that the nonreductive β-elimination worked (Fig. 4.108).

This is most likely a cleavage of the sulfate substitution of the free acid sulfate ester (Fig. 4.109).

This ion, m/z 377.3, occurs only in the unreduced β-eliminated O-linked oligosaccharide from bovine thyroglobulin. There are three

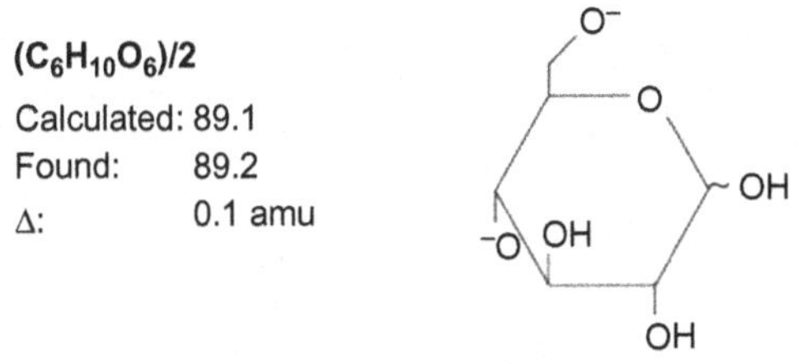

$(C_6H_{10}O_6)/2$
Calculated: 89.1
Found: 89.2
Δ: 0.1 amu
O⁻
O
OH
⁻O
OH
OH

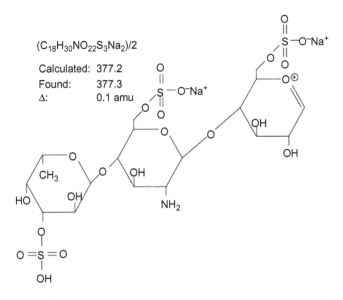

Figure 4.107 API MS, negative ion, of non-reductive β-elimination of bovine thyroglobulin O-linked oligosaccharide on API 2000.

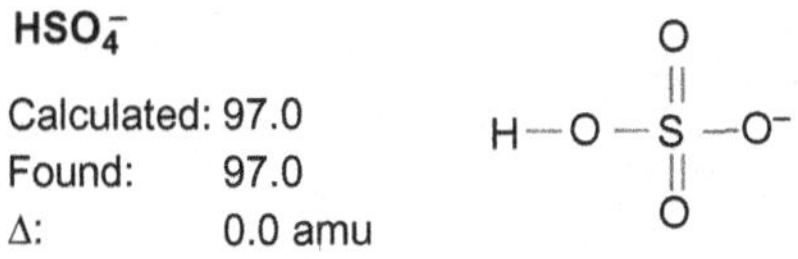

HSO_4^-
Calculated: 97.0
Found: 97.0
Δ: 0.0 amu
H—O—S—O⁻
O
O

Figure 4.108

$(C_{18}H_{30}NO_{22}S_3Na_2)/2$
Calculated: 377.2
Found: 377.3
Δ: 0.1 amu
O⁻Na⁺
O⁻Na⁺
CH3
HO
OH
OH
NH2
OH
OH
O=S=O
OH

Figure 4.109

sulfates, and it shows the loss of ketene common in the mass spectrometry of acetamido sugars such as glcNAc (Fig. 4.110).

This ion, m/z 179.2, is most likely a cleavage product of the reducing end of the sugar, sulfate hydrolyzed yielding glucose and sulfate m/z 97.1 (Fig. 4.111).

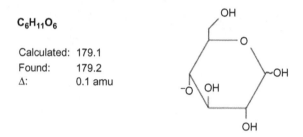

C₆H₁₁O₆

$C_6H_{11}O_6$

Calculated: 179.1
Found: 179.2
Δ: 0.1 amu

Figure 4.110

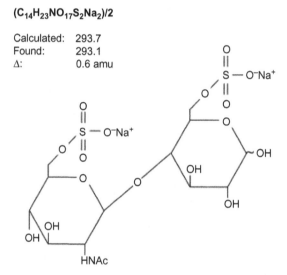

(C₁₄H₂₃NO₁₇S₂Na₂)/2

$(C_{14}H_{23}NO_{17}S_2Na_2)/2$

Calculated: 293.7
Found: 293.1
Δ: 0.6 amu

Figure 4.111

This ion, m/z 293.2, is found in the NRBE MS, but not in the RBE MS. This dianion is possibly glucose at the reducing sugar, GlcNHAc as the nonreducing end of this disaccharide, two primary position sodio-sulfate substitutions. This partially supports that the disulfated trisaccharide is not decomposed in our isolation process but loses fucose upon fragmentation in the mass spectrometer. Here is another example of sodium salts of sulfates, although technically neutral, producing one negative counter-charge for each sodium ion. We think sodium salts make sulfates more difficult to hydrolyze with base because the anionic charge of the sulfate is delocalized over three atoms, oxygen–sulfur–oxygen. The ammonium ion adds a

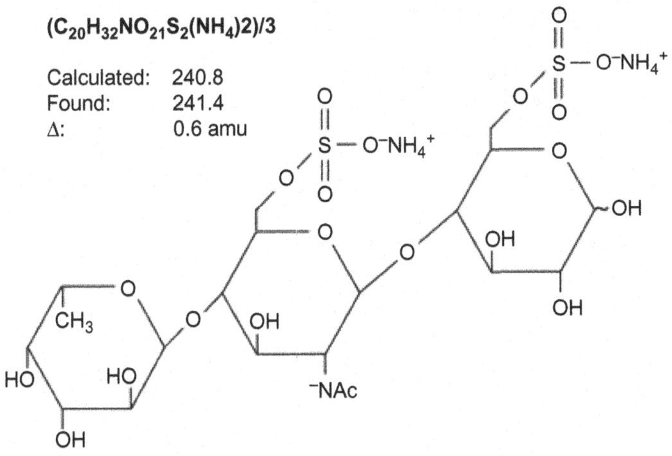

$(C_{20}H_{32}NO_{21}S_2(NH_4)2)/3$

Calculated: 240.8
Found: 241.4
Δ: 0.6 amu

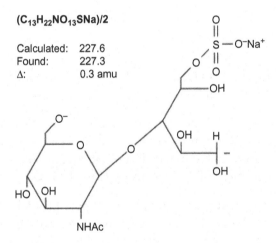

Figure 4.112

$(C_{13}H_{22}NO_{13}SNa)/2$

Calculated: 227.6
Found: 227.3
Δ: 0.3 amu

Figure 4.113

proton making the sulfate the free acid, more easily hydrolyzed (Fig. 4.112).

This trianion shows the intact trisaccharide as the diammonium salt acetamido anion. It is a direct indication that this process, nonreductive β-elimination of O-linked oligosaccharides from glycoprotein, gives undegraded intact unreduced oligosaccharide (Fig. 4.113).

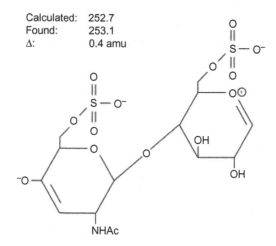

$(C_{14}H_{19}NO_{15}S_2)/2$

Calculated: 252.7
Found: 253.1
Δ: 0.4 amu

Figure 4.114

This is a cross-ring cleavage product for this compound from our new method. It shows three substitutions, the 6-sodio-sulfate, the GlcNAc substitution to glucose and the 6′ substitution, unknown from this structure but probably with sulfate.

This ion could be sulfated anhydro-fucose, m/z 227.1. Perhaps positive ion MS would discern them. With an oxonium ion at carbon one, for both unsulfated and sulfated, in the positive mode, the ions would be m/z 147.1 and m/z 227.1 respectively (Fig. 4.114).

This trianion positive oxonium ion is possibly dehydrated at the 3′,4′ positions and an oxonium ion, probably indicating reducing sugar at carbon 1 of glucose. Also indicated here is substitution on the glcNAc portion of the molecule and then cleavage there at the 4′ position of glcNAc. We've observed that sometimes counter-ions are included and sometimes not (Fig. 4.115).

This structure shows hexose substituted at the 4 and 6 positions, presumably sulfate at the 6 position and GlcNAc at the 4 position. Again this is an anion showing the integrity of the reducing sugar. We are unsure of why this is not an oxonium ion but this is the only structure that fits the m/z value that we could come up with (Fig. 4.116).

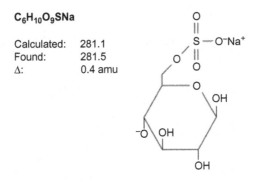

$C_6H_{10}O_9SNa$

Calculated: 281.1
Found: 281.5
Δ: 0.4 amu

Figure 4.115

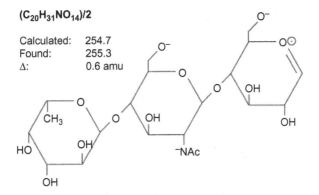

$(C_{20}H_{31}NO_{14})/2$

Calculated: 254.7
Found: 255.3
Δ: 0.6 amu

Figure 4.116

This trianion oxonium salt has a net two negative charge and is the carbohydrate backbone of the molecule. Two primary hydroxyl group substitutions are probably the sulfate esters (Fig. 4.117).

This dianion oxonium ion with sulfate substitution at the 6 position is common for reducing sugars. This hexose is glucose as shown by HPAEC-PAD. M/Z 241.4 has two possible structures, this structure and the structure noted in Fig. 4.27. The only way to discern the correct structure is through exact mass spectrometry or perhaps by MS/MS (Fig. 4.118).

This tetraanion (four negative charges) with one positive oxonium ion, with a net negative 3 charge, is probably the glucose and glcNAc portion of the molecule with two sodio-sulfate substitutions at primary hydroxyl positions. Fucose has been cleaved off. Under similar MS conditions, at times the oxonium ion is produced and sometimes not.

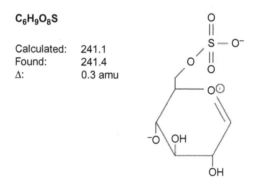

$C_6H_9O_8S$

Calculated: 241.1
Found: 241.4
Δ: 0.3 amu

Figure 4.117

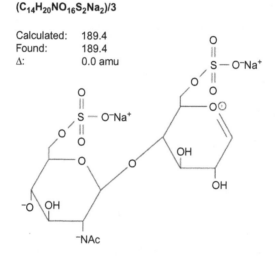

$(C_{14}H_{20}NO_{16}S_2Na_2)/3$

Calculated: 189.4
Found: 189.4
Δ: 0.0 amu

Figure 4.118

Here is the case where an oxonium ion is produced at the reducing end of the molecule (Fig. 4.119).

This is a tetraanion oxonium ion. It shows all of the substitution cleavages, noted here at the two primary hydroxyl positions, oxonium ion at C-1 of glucose and the four position of glcNAc. The acetamido proton is removed as well. The fucose of the trisaccharide is cleaved off (Fig. 4.120).

This anion oxonium ion gives credence to glucose as the reducing sugar since oxonium ions usually occur at the reducing end of reducing oligosaccharides. This molecule comes from the nonreducing

$(C_{14}H_{20}NO_{10})/3$

Calculated: 120.8
Found: 121.2
Δ: 0.4 amu

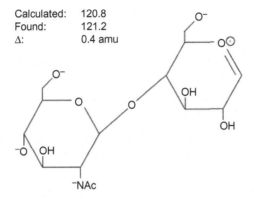

Figure 4.119

$C_6H_{10}O_8SNa$

Calculated: 265.1
Found: 265.3
Δ: 0.2 amu

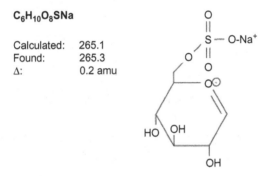

Figure 4.120

conditions of the analysis. Even though it has the same m/z value as an ion from the reductively β-eliminated O-linked oligosaccharide from bovine thyroglobulin, they have different structures because of the different chemistries performed on them.

4.17 MS Analysis of O-Linked Oligosaccharide Bovine Thyroglobulin First NRBE With NaOH, pH 11.4 Then RBE With pH 11.4 NH₄OH/NaBH₄[5] (Fig. 4.121)

This is an API MS, negative ion, nonreductive β-eliminated and reduced bovine thyroglobulin O-linked oligosaccharides. We call this "NRBE and RBE" as opposed to "RBE or NRBE". This spectrum is nearly identical to the reductive β-elimination produced spectra of this oligosaccharide. There are only two ions that are different as shown in Fig. 4.122.

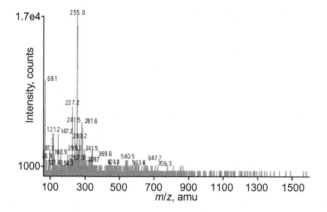

Figure 4.121 This is an API MS, negative ion, non-reductive β-eliminated and reduced bovine thyroglobulin O-linked oligosaccharides. We call this "NRBE and RBE" as opposed to "RBE or 'NRBE". This spectrum is nearly identical to the reductive β-elimination produced spectra of this oligosaccharide. There are only two ions that are different. We show them below.

(C₆H₁₃NO₇S)/2

Calculated:	121.6
Found:	121.2
Δ:	0.4 amu

Figure 4.122

(CH₆O₄SNa)/2

Calculated:	68.5
Found:	69.1
Δ:	0.6 amu

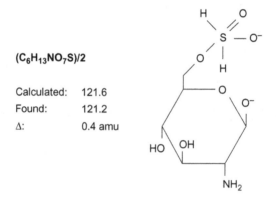

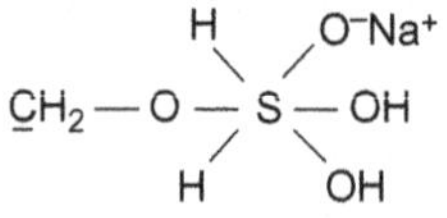

Figure 4.123

Here we have another example of reduced sulfate and glcNAc without acetyl group in this slide (Fig. 4.122).

This is direct evidence for the reduction of sulfate is a primary alcohol group of glucitol or glcNAc (Fig. 4.123).

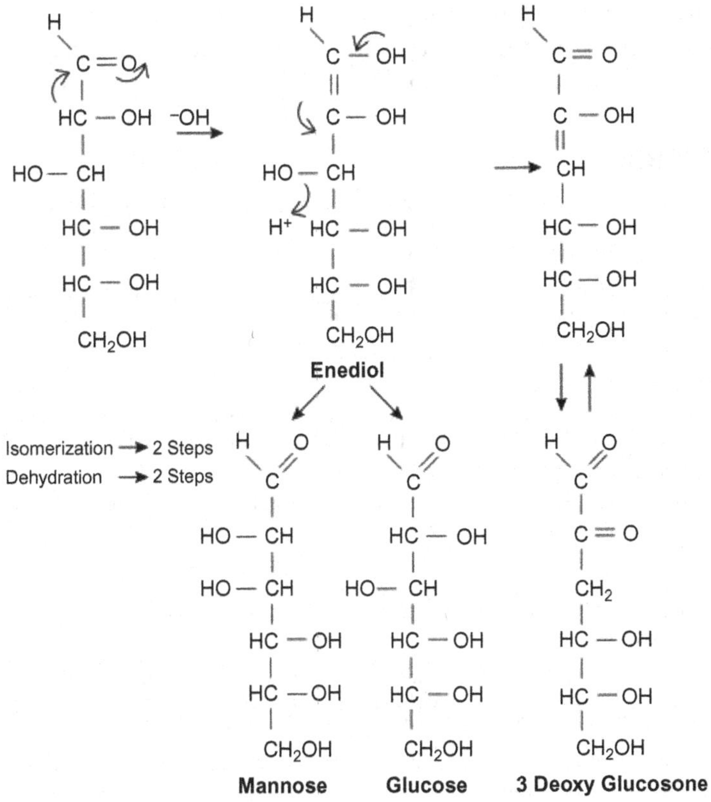

Figure 4.124

4.18 Mechanism for Acid/Base Catalyzed Isomerization and Dehydration of Glucose (Fig. 4.124)

This figure shows the mechanism for acid/base catalyzed isomerization and dehydration of glucose. If we take a solution of glucose and add pH 11.4 NaOH and allow it to stand less than 45–50 minutes (nonreductive β-elimination conditions) we find only glucose. But if we allow longer incubation times, such as 70 minutes standing at room temperature, we find appreciable quantities of mannose. The quantity of mannose increases with even longer times.

CONCLUSIONS

With the mass spectrometric techniques noted here and the ancillary techniques that give information critical to initiate and complete structural characterization of glycans, noted above, we are able to provide a structural profile that aids in the evaluation of the part that glycans play in disease mechanisms [43–52] and other applications.

REFERENCES

[1] Madson M, Mountain View CA. Methods of detecting N- and O-Linked oligosaccharides in glycoproteins. Sunnyvale, CA: Dionex Co. Patent application publication US20070105179A1 issued May 10, 2007.

[2] Madson M, Rao S, Avdalovic N, Pohl C. A simple procedure for the isolation of N- and O-linked oligosaccharides from glycoproteins. Glycobiology Meeting, November 2005, poster.

[3] Madson M. Method of testing of a milk trisaccharide. US patent issue US8993226B2 March 31, 2015.

[4] Madson M, Rao S, Avdalovic N, Pohl C. Structural analysis of oligosaccharides in skim milk and bovine thyroglobulin by HPLC and MS. Glycomics Meeting, San Diego, CA, March 2006, Poster presented.

[5] Madson M, Rao S, Avdalovic N, Pohl C. Structural analysis of oligosaccharides in milk and bovine thyroglobulin by HPLC and MS. International Carbohydrate Symposia, Whistler, BC, Canada, August 2006, Poster presented.

[6] Madson M, Zhang Z, Rao S, Slingsby R, Pohl C. Structural analysis of oligosaccharides in milk and bovine thyroglobulin by HPLC and MS. Glycobiology Meeting November 2006. Poster presented.

[7] Christus J, Madson M. Simple method for the non-reductive β-elimination of O-linked oligosaccharides of glycoproteins biologistics LLC Garner, IA 50438 USA which was given as an oral presentation to the 245th ACS Meeting, New Orleans, LA; 2013.

[8] Christus J, Madson M. Structural characterization of a new oligosaccharide from banana fruit ethanol extract biologistics LLC Ames, IA 50010 249th ACS Meeting, Denver, CO, March 24, 2015.

[9] Der Agopia RG, Purgatti E, Cordenunsi BR, Loyalo FM. Synthesis of fructooligosaccharides in banana 'prata' and its relation to Invertase activity and sucrose accumulation. J? Agric Food Chem 2009;57(22):10765–71.

[10] Allen PJ, Bacon JS. Oligosaccharides and associated glycosidases in Aspen tissue. Biochem J 1956;63(2):200–6.

[11] Bacon JSD. The trisaccharide fraction of some monocotyledons. Biochem J 1959;73 (3):507–14.

[12] Bacon JSD, Edelman J. The carbohydrates of Jerusalem artichoke heart and other compositae. Biochem J 1951;48(1):114–26.

[13] Forsythe KL, Feather MS. The detection of isokestose and neokestose in plant extracts by ^{13}CNMR. Carbohydr Res 1989;185:315–19.

[14] Forsythe KL, Feather MS, Gracz H, Wong TL. Detection of kestosesand neokestose-related oligosaccharides in extracts of Fatsuca arundinacea, Dactylls glomerata and Asparagus officinatis cultures and Invertase by ^{13}C and ^{1}H nuclear magnetic resonance spectroscopy. Plant Physiol 1990;92:1014–20.

[15] Fuller KW, Northcote DH. A micromethod for the separation and determination of polysaccharides by zone electrophoresis. Biochem J 1956;64(4):657–63.

[16] Killian S, Kritzinger S, Rycroft C, Gibson G, du Preez J. The effect of novel bifidogenic trisaccharide, neokestose, on the human colonic microbiota world. J Microbiol Biotechnol 2002;18(7):637–44.

[17] Liu JH, Waterhouse AL, Chatterton NJ. Proton and carbon chemical shift assignments for 6-kestose and neokestose. Carbohydr Res 1991;18(217):43–9.

[18] Shiomi N. Isolation and identification of 1-kestodse and neokestose from onion bulbs. J? Faculty Agriculture-Hokkaido University 1978;58(4):548–56.

[19] Suzuki T, Maeda T, Grant S, Grant G, Sporn P. Confirmation of fructose biosynthesized from [1-^{13}C] glucose in asparagus tissue using MALDI-TOF-MS and ESI-MS. J Plant Physiol 2013;170(8):715–22.

[20] Christus J, Madson M. Structural identification of an unknown trisaccharide in bovine milk. 243rd National American Chemical Society Meeting, San Diego, CA, March 2012, Oral Presentation.

[21] Christus J, Madson M. Isolation of amino acid-linked O-linked oligosaccharides from K-casein and bovine submaxillary mucin BioLogistics LLC, IA 50010 oral presentation to ACS, submitted for ACS National Meeting, Boston, MA, Fall 2015.

[22] Madson M, Christus J. Structural characterization of an unknown di-phosphorylated bovine submaxillary mucin O-linked oligosaccharide biologistics LLC Ames, IA 50438 Poster presented to 247th ACS Meeting, Dallas, TX, Spring 2014.

[23] Saito T, Itoh T. Variations and distributions of O-glycosidically linked sugar chains in bovine K-casein. J Dairy Sci 1992;75(7):1768–74.

[24] Christus J, Madson M. Structural characterization of a compound from the ethanol extract of banana fruit, 249th National American Chemical Society Meeting, Denver, CO, March 2015, Accepted Poster.

[25] Madson M, Christus J. MS method to discern phosphate versus sulfate esters of carbohydrates biologistics LLC Garner, IA 50438 given as an oral presentation to 246th ACS Meeting, Indianapolis, IN; 2013.

[26] Madson M. MS method for discernment of sulfate versus phosphate carbohydrate esters. Patent application publication 2015 US20150064793A1.

[27] Mawhinney T, Adelstein E, Morris D, Mawhinney A, Barbero G. Structure determination of five sulfated oligosaccharides derived from tracheobronchial mucus glycoproteins. J Biol Chem 1987;262(7):2994–3001.

[28] Chance DL, Mawhinney T. Disulfated oligosaccharides derived from tracheobronchial mucous glycoproteins of a patient suffering from cystic fibrosis. Carbohydr Res 1996;295:157–77.

[29] Thomsson K, Karlsson N, Hansson G. Liquid chromatography-electrospray mass spectrometry as a tool for the analysis of sulfated oligosaccharides from mucin glycoproteins. J Chromatograp A 1999;854:131–9.

[30] Yuen CT, Bezouska K, O'Brien J, Lemoine R, Lubineau A, et al. Sulfated blood group Lewis(a). A superior oligosaccharide ligand for human E-selectin. J Biol Chem 1994;269 (3):1595–8.

[31] Bubb WA, Urashima T, Kosho K, Nakamura T, Arai I, Saito T. Occurrence of an unusual lactose sulfate in Dog Milk. Carbohydr Res 1999;318:123–8.

[32] Parkkinen J, Finne J. Isolation of sialyl oligosaccharide phosphates from bovine colostrum and human urine. Methods Enzymol 1987;138:289–300.

[33] Madson, M. Method for Testing a Milk Trisaccharide. US patent issued 3/2015 US8996226B2.

[34] Parkkinen J, Finne J. Occurrence of N-acetyl glucosamine-1-phosphate in complex carbohydrates. Characterization of phosphorylated sialyl oligosaccharide from bovine colostrum. J? Biol Chem 1985;260:10971–5.

[35] Gopal P, Gill H. Oligosaccharides and glycoconjugates in bovine milk and colostrum. Br J Nutr 2008;84(Suppl. 1):S69–74.

[36] Pan GG, Melton LD. Lactones of disialyl lactose; characterization by NMR and mass spectra. Carbohydr Res 2006;341:730–7.

[37] Hu F, Furihata K, Ito-Ishida M, Kaminogawa S, Tanokura M. Nondestructive observation of bovine milk by NMR spectroscopy: analysis of existing states of compounds and detection of new compounds. J Agric Food Chem 2004;52:4969–74.

[38] Patel T, Parekh R. Release of oligosaccharides from glycoproteins by hydrazinolysis. Meth Enzymol 1994;230:57–66.

[39] Tsuiki S, Hashimoto Y, Pigman W. Comparison of procedures for the isolation of bovine submaxillary mucin. J Biol Chem 1961;236:2172–8.

[40] Mechref Y, Novotny M. Structural studies of gycoconjugates at high sensitivity. Chem Rev 2002;102(2):321–70.

[41] Halima A, Brinkmalmb G, Ruetschia U, Westman-Brinkmalmb A, Porteliusb E, Zetterbergb H, et al. Site-specific characterization of threonine, serine, and tyrosine glycosylations of amyloid precursor protein/amyloid β-peptides in human cerebrospinal fluid. Proc Natl Acad Sci USA 2011;108(29):11848–53.

[42] Hill H, Reynolds J, Hill R. Purification, composition, molecular weight and subunit structure of ovine submaxillary mucin. J Biol Chem 1977;252:3971–8.

[43] Bode L, Muly-Reinholz M, Mayer K, Seeger W, Rudolf S. Inhibition of monocyte, lymphocyte and neutrophil adhesion to endothelial cells by human milk oligosaccharides. Thromb Haemost 2004;92(6):1402–10.

[44] Maury LP, Teppo AM, Wegelius O. Relationship between urinary sialylated saccharides, serum amyloid A protein, C reactive protein in rheumatoid arthritis and systematic lupus erythematous. Ann Rheum Dis 1982;268–71.

[45] Masuda M, Kawase Y, Kawase M. Enzymic synthesis of α (2-3) sialyl lactose using a membrane reactor. Seibutsu Kogaka Kaishi 2001;79(9):345–8.

[46] Otek I, Hasty DC, Sharon N. Antiadhesion therapy of bacterial diseases: prospects and problems. FEMS Immunol Med Microbiol 2003;38:181–91.

[47] Kijima-Suda I, Miyamoto Y, Itoh M, S. Toyoshima S, Osawa T. Possible mechanism of inhibition of experimental pulmonary metastasis of mouse colon adenocarcinoma 26 sublines by a sialic: nucleoside conjugate. Cancer Res 1988;8(13):3728–32.

[48] Zou Z, Chastain A, Moir S, Ford J, Trandem K, Martinelli E, et al. Siglecs facilitate HIV-1 infection of macrophages through adhesion with viral sialic acids. PLoS One 2011;6(9): e24559.

[49] Orllandi PA, Klotz FW, Haynes JP. A malaria invasion receptor, the 175 kilodalton erythrocyte binding antigen of plasmodium falciparum recognizes the terminal Neu5Ac (α2-3) gal- sequences of glycophorin A. J Cell Biol 1992;116(4):901–9.

[50] M'rabat L, van Baalen A, Stahl B, Vos AP, Snoeren THM. Milk oligosaccharide for stimulating the immune system. Application number: 12/29468, Publication date: November 2, 2010. Assignee: N.V. Nutricia.

[51] Cardenas Delgado VM, Nugnes LG, Colombo LL, Troncoso MF, Fernandez MM, Malchiodi El, et al. Modulation of endothelial cell migration and angiogenesis; novel function for the tandem repeat lectin galectin-8. FASEB J 2011;25:242–54.

[52] Y-Moriguchi T, Yu L, Stalnaker H, Davis S, Kunz S, Madson MA, et al. O-mannosyl phosphorylation of α dystroglycan is required for Laminin binding. Science 2010;327 (5961):88–92.

Printed in the United States
By Bookmasters